Concentration of some commercially available Acids and Alkalies

	%	specific gravity	concentration [mol l^{-1}]
HCl	25	1·127	7·7
	36–38	1·18	12
HNO_3	65	1·40	14
	70	1·42	16
	96	1·50	23
	100	1·52	24
H_2SO_4	25	1·36	3
	96–98	1·84	18
H_3PO_4	88–90	1·75	16
	99	1·88	19
NH_4OH	25	0·91	13·4
	30	0·896	16
	35	0·88	19·5

Solubility Products

	18°C	25°C	100°C
AgCl	1×10^{-10}	2×10^{-10}	$2{\cdot}5 \times 10^{-10}$
$BaCO_3$	7×10^{-9}	8×10^{-9}	
$BaSO_4$	$0{\cdot}9 \times 10^{-10}$	1×10^{-10}	$2{\cdot}6 \times 10^{-10}$
$CaCO_3$	1×10^{-8}	1×10^{-8}	
CdS	5×10^{-29}		
FeS	4×10^{-19}		
$PbSO_4$	1×10^{-8}		
ZnS	1×10^{-23}	1×10^{-24}	

Though this be madness, yet there is method in't.

Hamlet, II, ii (211)

METHODS FOR PHYSICAL AND CHEMICAL ANALYSIS OF FRESH WATERS

IBP HANDBOOK No 8

Methods for Physical and Chemical Analysis of Fresh Waters

by

H. L. GOLTERMAN

Limnological Institute
Nieuwersluis
Netherlands

R. S. CLYMO

Westfield College
London, NW3 7ST

M. A. M. OHNSTAD

Freshwater Biological Association
Ferry House
Ambleside
Cumbria, U.K.

Second Edition

BLACKWELL SCIENTIFIC PUBLICATIONS
OXFORD EDINBURGH LONDON MELBOURNE

Osney Mead, Oxford, OX2 0EL
8 John Street, London, WC1N 2ES
9 Forrest Road, Edinburgh, EH1 2QH
P.O. Box 9, North Balwyn, Victoria, Australia.

First published 1969
Second edition 1978

British Library Cataloguing in Publication Data
Golterman, Hendrik Leonard
Methods for physical and chemical analysis of fresh waters.—2nd ed.—(International Biological Programme. Handbooks; no. 8).
1. Fresh water—analysis
I. Title II. Clymo, Richard Stringer III. Ohnstad, M A M
IV Series
546'22 QD142

ISBN 0-632-00459-2

Distributed in the United States of America by
J.B. Lippincott Company, Philadelphia,
and in Canada by
J.B. Lippincott Company of Canada Ltd, Toronto.

Typeset by Enset Limited, Midsomer Norton, Avon

Printed in Great Britain by
Billing & Sons Limited
Guildford, London and Worcester
and bound by
Mansell (Bookbinders) Ltd., Witham, Essex

CONTENTS

*Method deleted from this edition.

*Method deleted from this edition.

*Method deleted from this edition.

8 DISSOLVED GASES; O_2, N_2, CO_2, H_2, CH_4

*Method deleted from this edition.

PREFACE

Since the first edition was published in 1969 the IBP has ended but the demand for this manual persists, partly because more people are making chemical analyses of fresh waters. Encouraged by colleagues and publisher, we hope that this edition will continue to meet the need for a simple relatively cheap source of advice about chemical analyses of freshwaters.

We have added a section about measurement of light and temperature, because people who need chemical analyses very often need to know about light and temperature too, and measure them at the same time that they take samples for chemical analysis.

The explanatory sections of the first edition seem to have been of unexpected value particularly to graduates in biology. We have therefore, increased the amount of explanation in the Introduction. The use of the book for day to day reference should not be affected. But the Introduction does contain matter which is of importance to all users of the book, and we hope all will read it.

To keep the book short we have had to remove some methods which have proved to be little used or unreliable. The description of others has been shortened. To preserve compatibility we have kept the original method numbers. Methods deleted from this edition are listed in the Contents however so that the user needing them should be able to trace them.

We have had difficulty in deciding which units to use. The SI system is gradually gaining acceptance, but there does seem to be a lot of resistance to it. Textbooks may aim to change such attitudes but because this book is of a practical nature we have compromised and kept the old style volumetric unit 'l' instead of the SI unit 'dm^3', and retained M as a convenient unit of chemical concentration ($M = mol\ dm^{-3} = mol\ l^{-1}$), despite the view of McGlashan (1968) that the use of such units '. . . is to be progressively discouraged and eventually abandoned'. By 1971 McGlashan was able to report that the litre had official recognition as a nickname and 'l' as a nicksymbol within the SI system. We have also kept '°C' for temperature measurement instead of the SI unit 'K'.

On the other hand there are increasing numbers of biologists who have never come across the *concepts* of 'normality' or of 'equivalence' and to whom '$1{\cdot}2\ meq\ l^{-1}$' or '5 N' are meaningless. We therefore express chemical amounts in the unit 'mol' and specify the chemical species intended. Further details are given in §1.2.2.

For the same reason we use 'absorbance' in place of 'extinction' and 'optical density'.

We realise this compromise will annoy many people but believe it will minimise difficulties. Fortunately the eternal principles of poise and equipoise ensure that those who are most annoyed are those least easily confused.

The first edition was based in part on reports by chairmen of working groups of an IBP symposium 'Chemical Environment in the Aquatic Habitat' held in Amsterdam and Nieuwersluis in Oct 1966. These chairmen, and the chapters in which their reports were incorporated, were: F.A.Armstrong (5), A.F.Carlucci (7), E.K.Duursma (5, 7), J.E.Hobbie (7), C.J.Hogendijk (4), A.V.Holden (7), N.J. Nicolson (3), H.A.C.Montgomery (8), S.Olsen (5), D.Povoledo (7), A.Rebsdorf (3), J.Shapiro (6), S.Yentsch (3, 7).

After the third printing of the first edition a small 'workshop' meeting considered what changes should be made. Those who contributed to this meeting were S.E.Allen, F.A.J.Armstrong, E.K.Duursma, H.A.C.Montgomery, and D.Povoledo. If merit there be in this edition they should share it, but they are in no way responsible for defects.

Dr. D.Westlake, Dr. J.F.Talling, Dr. A.V.Holden, Mr. J.Voerman and Mr. M.French have made most useful criticisms of several parts of the book.

Mr. J.Voerman and Dr. I.Butcher have converted illegible manuscript into typescript. The librarians of the FBA have located even the most obscure references.

Many other colleagues have helped by informal discussion.

To all we give our thanks.

We also thank Robert Campbell of Blackwell Scientific Publications for his tolerance of our asynchronous lives, and Nel, Kay and Ray for their continuing advice and forbearance.

Finally we thank Dr. J. Rzoska who had such an important influence on the writing of all the IBP (PF) manuals.

The first edition had one editor and an assistant. This edition has a different name and three authors. We fear that librarians will be confused but trust in the hope that the book will spend more time on laboratory benches than on library shelves.

CHAPTER 1

INTRODUCTION

1.0 SCOPE AND LIMITATIONS OF THIS MANUAL

When separate workers study the same problem it is useful for them to use at least some of the same methods in order to allow direct comparison of their results.

This manual describes some methods for chemical analyses of fresh waters. We have chosen those analytes which are most often of importance to workers interested in the biological processes which lead, with interconversion of energy, to the production of organic material. The chemical environment has important effects on such processes. Exactly how is a matter of interpretation and is not considered here.

Most of the methods are intended for filtered water only. Some may be used, with the modifications described, for particulate matter too. A few are intended for particulate matter only.

Sufficient detail and explanation have been included to make the main part of the manual useful to those with little training in either chemistry or limnology. Chapter 9 is an elementary description of the physical chemistry of dilute solutions.

Somc other books about chemical analysis or limnochemistry are listed with a prefatory asterisk (*) in the reference list at the end of the book.

The user of the methods described here should be aware of a number of difficulties.

First is the fact that fresh waters are in a chemically delicate state but analyses must often be made in strongly acid or alkaline conditions, and the state of the compound studied is thereby changed. For example during phosphate analysis hydrolysable compounds are converted to H_3PO_4 by the acid necessary for the determination (Olsen, 1967). Such changes may in some cases be of biological importance and in others not.

Secondly, the method may not be as specific as is supposed. Titration with silver nitrate is usually assumed to estimate chloride, but if iodide is present then it will be included in the estimate. The limnochemist must be constantly watchful for such possibilities.

Thirdly, some small or large fraction of the substance estimated by a chemical analysis may not be available to one or more of the organisms of interest. The only solution to such problems is to use bioassay, in which the organism itself is used to estimate the 'available' concentration. Such assays may be the *only* method of detecting the minute concentration of vitamins. But there are great difficulties, in theory and in practice, in using bioassays. Such details are beyond the scope of this book. So too are the other forms of bioassay: many aquatic organisms from algae to fish produce and release chemicals (pheromones) that can cause behavioural responses by other members of the same species or by different species

(Vallentyne, 1967). The behaviour is the assay. This type of work can be done with simple apparatus and could be of great economic importance.

There is a serious danger that because a method appears in this book the reader may assume that meaningful numbers will always result if the directions are followed. This is not so. The individual worker has the responsibility for deciding if a method given in this manual is doing the job he or she requires.

There is a second danger. Methods are not ideal and sometimes a completely new approach is necessary. The very existence of a manual may inhibit such work, important and necessary though it is.

1.1 LEVELS OF ANALYSIS

The limnochemist's working conditions vary from solitary fieldwork (perhaps in a remote area) to a well equipped laboratory with abundant expertise to hand. The methods in this manual have therefore been described at one or more of three levels of refinement. These correspond to differences in the available equipment and exactitude, though the different levels are often variations on the same theme.

1.1.1 Level I

At level I are methods which can be used in the field or in places where only primitive laboratory facilities are available. In the extreme case the limnologist arrives at the lake, laboratory in rucksack. Constituents such as oxygen, chloride, bicarbonate and calcium may all be measured by titration using a simple calibrated pipette to an accuracy of about $\pm 10\%$. Similar accuracy is obtained using a plastic syringe and counting the drops as they fall into a clear plastic tube which acts as titration vessel. This system is compact, light, nearly indestructible and can be used in a boat (Denny, 1977).

Nowadays the limnologist often needs to measure other biologically important constituents too. Amongst these are phosphate, nitrate and silicate for which colorimetric methods are usually necessary. Accuracy of perhaps $\pm 20\%$ may be obtained by matching the colour produced by the sample against standards having colours which bracket the sample. Greater accuracy may be obtained with a comparator which contains a set of tinted glasses. A well known one is made by Lovibond. It occupies about 0·5 dm^3 and is easily carried. More cumbersome is a battery operated colorimeter. A simple one is made by *Dr. Lange. The accuracy obtainable is not of course as great as that with a spectrophotometer, but is adequate for level I methods.

Portable instruments for the specific measurement of pH, conductivity and O_2 concentration are now reliable provided that their 'electrode' systems are carefully treated. It is arguable that if only three measurements can be made then these should be of pH, conductivity and O_2 concentration because these are poorly correlated and taken together give a good idea of the general type of water.

It should be remembered that level I methods are just as liable to chemical interference as are the more complicated level II and III methods. The danger is the greater because level I methods are usually used on water of unknown character.

The results of level I methods nearly always lead to questions—for example about daily, seasonal, or spatial variation—which can be answered only by the more accurate methods of level II.

* Dr. Lange GmbH, Hermanstrasse 14, Berlin 37, Germany.

1.1.2 Level II

At level II the methods can be applied in a moderately well equipped laboratory. Most methods have as the final step a volumetric, colorimetric or simple flame emission colorimetric determination. The user of level II methods is assumed to know the basic facts—both theoretical and practical—of analytic chemistry. In particular the user should know the Beer and Lambert laws and be able to make acid-base or $KMnO_4$ titrations with an accuracy of $\pm 0{\cdot}2\%$.

The laboratory must contain a balance (accuracy ± 0·1 mg) and a piston microburette (accuracy ± 0·001 ml), a spectrophotometer or colorimeter working in the range 380 to 800 nm, and a flame emission colorimeter. It should be possible to make colorimetric measurements in the absorbance range 0·05 to 0·6 with an accuracy and precision of three significant figures.

Colorimeter filters should be of narrow bandwidth. Transmittance at 15 to 20 nm on either side of the wavelength of maximum transmittance should be less than 50% of the maximum. Spectrophotometer wavelength settings should be stable and reproducible with no detectable backlash. With colorimeter and spectrophotometer it should be possible to go through the sequence 'dark' reading, 'blank' reading at least three times with no detectable change in corresponding readings.

A stock of borosilicate glassware is also essential.

If more samples are entering the laboratory than can be analysed the limnochemist should consider first: 'Is it really necessary to analyse so many samples?' If it is necessary then he should consider the possibilities of level III methods.

1.1.3 Level III

At this level two types of methods are placed. First there are determinations which are not routine in most laboratories. Examples are analyses of organic carbon, some organic compounds and of trace elements (for which the making of valid 'blanks' is not easy). These methods need special equipment and, often, a lot of experience with microanalytical techniques. The user of such methods must expect to spend days or weeks in mastering the technique using known standards.

Secondly are methods which are essentially at level II but in which part of the analysis is made by a complex, often automatic or semi-automatic, machine. Detailed instructions about the use of such machines are always supplied by the manufacturers and advances in this field are rapid, so this manual indicates only that a method can be automated. A review of automatic methods is given by Lee (1967). Auto-analysers are an example of such machines and with these the methods themselves must often be modified in ways too specific to the machine and too numerous to be given here. We except atomic absorption flame spectrophotometers however, and give some details (§ 1.4.3).

It is sometimes mistakenly supposed that level III methods are unsuitable for fieldwork but it is, for example, quicker and more informative to make dissolved oxygen measurements with an oxygen 'electrode' connected to a recorder (§ 8.1.3) than by Winkler titration (§ 8.1.2).

Another mistaken belief is that when a level III method is introduced then the chemist becomes redundant. Without adequate checking automatic equipment will sooner or later begin to produce results of declining accuracy, and interferences in water samples of unknown character will go unrecognised. The knowledge of the chemist is just as necessary as it is with manual methods and he must in addition understand the priniciple on which the automatic machine works and be able to detect faults, particularly those which affect the results, even though the machine appears to be working properly.

The accuracy of properly functioning automatic equipment is generally the same as at level II because the methods are generally the same but it may be greater because machines do not become tired. The variability of replicates should be smaller than the equivalent manual method.

There are two main problems with automatic machines. First is the cost of, and delay in, correction of faults. This may be particularly serious if no competent assistance is available in the laboratory so that reliance must be placed on the manufacturer or his agents.

Secondly, it becomes all too easy to collect a mass of data, so that the reason for doing so is not considered. An automatic machine can easily become a Sorcerer's Apprentice. Even worse, the analyst may become a slave of the machine. These dangers are widely known but too little appreciated.

1.2 GENERAL

1.2.1 Errors, accuracy, precision, sensitivity, range.

The meaning of these terms varies with the user. We explain our intention here.

It is convenient to recognise three types of error: erratic mistakes, systematic errors, and random errors.

(i) Mistakes may be instrumental or human. Sometimes they can be definitely identified: for example the apparent duplication of a flask number. The flasks already done can be examined and the mistake can then be corrected. More commonly there is simply a strong suspicion that a mistake has been made: for example suppose that replicates are 5·4, 5·3, 5·4, 6·4, 5·2, 5·2. Common sense may reject 6·4 and it is possible to bolster the rejection by some quasi-statistical argument, for example the Chauvenet criterion. But observations of great interest may thereby be suppressed. The only entirely satisfactory solution is to avoid mistakes!

(ii) Systematic errors are related to the concepts of bias and accuracy, whilst random errors are related to those of precision, reproducibility and dispersion. In rifle shooting it is possible to get several shots grouped close together but a long way from the centre of the target. Such a group has high precision or reproducibility (small dispersion) with small random errors. But the same group has low accuracy because the mean is a long way from the true mean. The systematic error or bias is large. It is essential to recognise that *high precision may commonly accompany low accuracy*. The reverse—low precision but high accuracy—is more rare and would be more obviously unsatisfactory.

A lot is known about the treatment of random errors, so they tend to receive disproportionate attention. Systematic errors can be much more serious: for example the standard solution which is ten times too concentrated or, even worse because less likely to be noticed, 1·3 times too concentrated.

The simplest way of locating systematic error is to use two entirely independent methods based on different principles and using different standards. If the methods give the same result then they provide mutual support.

A common cause of systematic error in chemical analyses is chemical interference. The analytic reaction is affected by components of the sample besides the analyte. For example, the flame emission of K^+ is reduced by even small concentrations of PO_4. This type of systematic error may usually but not always be detected by the method of 'standard addition' (§ 1.2.3).

(iii) Even a modest analysis of random error is beyond our scope but the subject is important because in many analyses the random error is bigger, perhaps much bigger, than the analyst realises. We therefore give here a simple guide.

For illustration, consider the length of sentences in Charles Darwin's The Origin of Species (1859). The character or variable to be examined is the number of words in sentences and the 'universe of discourse' or population is finite in this case, being limited to this one book. The distribution of sentence length is shown in Fig. 1.1. It will be obvious that we could calculate the arithmetic mean length of sentences, μ, by taking the sum of individual sentence lengths, x, and dividing by the number of sentences, N.

$$\mu = \Sigma x/N$$

(There are other measures of centrality of which the median and mode are the most useful, but we ignore them here).

Another book might have the same mean length of sentence but a greater proportion of shorter and longer sentences. A convenient measure of the spread of lengths is given by the standard deviation, σ, where

$$\sigma^2 = \Sigma(x - \mu)^2/N = \text{the population variance}$$

The standard deviation has the same dimensions as the original measurements.

In this example it was practicable to enumerate the whole population and measure μ and σ exactly, but such cases are unusual. More often one must resort to 'sampling', that is to measuring only part of the population. One might for example take only 200 sentences from the book, or 27 bottles of 500 ml from a lake of area 1 km^2 and of depth 2 m. In such cases one can still calculate the mean and standard deviation (or its square, the variance) but these now refer to the sample not to the population. The distinction is shown by using roman letters:

$$\bar{x} = \Sigma x/n$$
$$s^2 = \Sigma(x - \bar{x})^2/(n - 1)$$

where n is the number of items in the sample. In practice $\Sigma(x - \bar{x})^2$ is awkward to calculate and it is usual to use the fact that

$$\Sigma(x - \bar{x})^2 = \Sigma x^2 - (\Sigma x)^2/n$$

Note that strictly the 27 bottles together constitute the sample, though it is common practice (which we follow) to call each bottle 'a sample'. If the sample contains more than about 20 items then the standard deviation will be nearly independent of sample size.

Generally we want to know about the population mean μ. The best estimate we have is $\bar{x}$, but it is only rarely and by chance that $\mu = \bar{x}$ (or that $\sigma = s$). We need therefore to discover how $\bar{x}$ for a series of samples is distributed about μ. Fig. 1.1 shows the values $\bar{x}$ of a large number of random samples of sentences. It is important to note that the samples are random: every possible item is given an equal chance of being chosen, or, seen in reverse, no bias is used in choosing samples. This may be difficult to arrange in practice perhaps, for example, because it is easier to sample near the shore of a lake than from its centre. But unless sampling is random then most of the common statistical procedures are invalid. The distribution of sample means is a remarkable one. For samples of more than about 20 it is approximately symmetrical with a peak close to the population mean μ, and a shape close to the 'normal' or 'Gaussian' distribution, even though the population

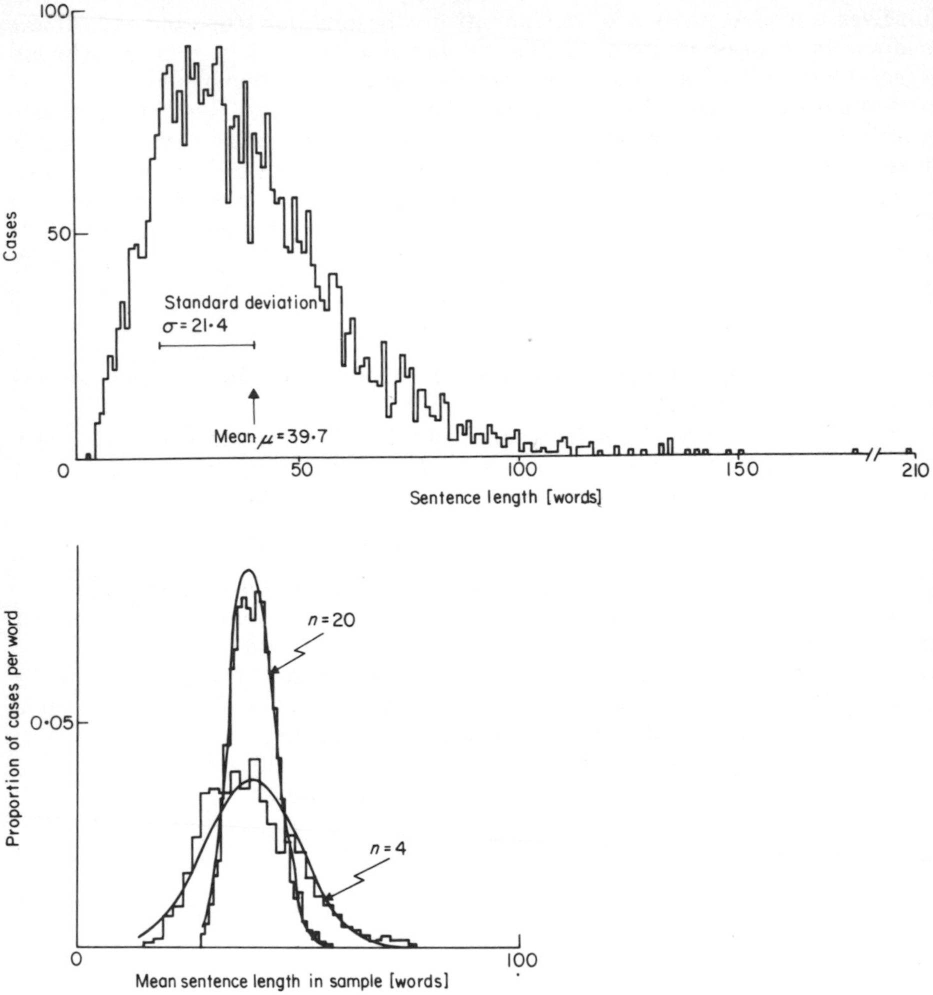

Figure 1.1

Top. Distribution of sentence length in 'The Origin of Species' (Darwin, 1859). This is the whole population of 3760 sentences. It has mean (μ) = 39·7 and standard deviation (σ) = 21·4 words per sentence. The shortest sentence is 3 words, the longest 208. The distribution is markedly lop-sided ('skewed').

Bottom. Distribution of means of 2500 samples from this population, for samples containing 4 sentences and for 20 sentences, chosen at random. The 'normal' curve with the same mean and standard error (= standard deviation of the sample distribution) is shown superimposed. For n = 20 the two coincide fairly closely but for n = 4 the sample distribution is skewed. For such small samples Student's 't' distribution would be more suitable. In both cases the mean for 2500 random samples is close to the population mean.

is markedly different. It is this fact—that the distribution of sample means is approximately normally distributed about the population mean—that makes the normal distribution so important. The standard deviation of this sampling distribution is called the standard deviation of the mean or standard error of the mean or simply standard error, $s_{\bar{x}}$. It is easily calculated from the sample variance, using: $$s_{\bar{x}}^2 = s^2/n$$

Unlike the sample standard deviation the standard error decreases as n increases: the more items in the sample the more likely it is that $\bar{x}$ is close to μ.

As the number of items in each sample increases so the sampling distribution becomes closer to the 'normal' distribution. A sample with more than 20 times may be assumed to come from a 'normal' distribution, but if there are fewer than 20 items then an appropriate 'small sample' distribution, for example 'Student's 't'', should be used.

Some authors use 's', the standard deviation, as a measure of precision, others use '$s_{\bar{x}}$', the standard error. If '$s_{\bar{x}}$' is used then 'n', the number of items, should be reported too. The value of 's' should be based on at least 20 items. On the whole 's' is to be preferred because it is an intrinsic property, little dependent upon 'n'. But if statistical tests are to be made to compare one mean with another then '$s_{\bar{x}}$' will have to be calculated. In this manual we use 's' as the measure of precision, with the exception of the next section.

One very useful property of the normal distribution is that if two such distributions are combined then the variance of the combined distribution is simply the independent sum of the separate variances. We use this fact to deal now with the accumulation of random errors during the course of an analysis.

Every measurement has random error associated with it. In a chemical analysis there may commonly be 5 to 20 separate measurements of volumes, absorbances, and so on, each with random error. All these random errors contribute to the overall random error, but in differing proportions. Take for example a simple colorimetric determination. A volume of sample U has reagents added to it. Assume that in this example the volume of these is not critical (though this is not always the case). The mixture is made up to a known volume V. The absorbance A is measured and compared with a straight calibration line made with standards subject to the same errors. The equation of the calibration line is $y = bx + a$ (§1.2.3) where y is the concentration of analyte in the volume V and x is the corresponding absorbance. The sample concentration, C, is then given by:

$$C = \frac{V}{U}(bA + a)$$

The usual approach to estimating the random error in C (and hence the precision) would be to make 20 or more replicate measurements and calculate s or $s_{\bar{c}}$ from these directly. This assumes that the random errors in U, V and A are being combined and measured directly, *but it also assumes that b and a are error free.* This is obviously untrue because both were estimated from a standard calibration line prepared in the same way as the samples. Suppose that we estimate the standard error of b and a as shown in § 1.2.3 and that we estimate the standard error of U, V and A by making replicate pipettings, and so on.

The variance of the mean of C is simply the square of the standard error of C. This variance is the sum of the variances contributed by each measurement. These variances in turn are the square of the standard error of each measurement weighted (multiplied) by a measure of its importance in the calculation. This measure, finally, is the partial differential coefficient, which is got by treating the equation as if it had one independent variable only.

For example the partial differential of C with respect to U is:

$$\frac{\partial C}{\partial U} = -\frac{V(bA + a)}{U^2}$$

The full equation for the variance of the mean of C is

$$s_{\bar{C}}^2 = \left(\frac{\partial C}{\partial V} s_{\bar{V}}\right)^2 + \left(\frac{\partial C}{\partial U} s_{\bar{U}}\right)^2 + \left(\frac{\partial C}{\partial A} s_{\bar{A}}\right)^2 + \left(\frac{\partial C}{\partial a} s_{\bar{a}}\right)^2 + \left(\frac{\partial C}{\partial b} s_{\bar{b}}\right)^2$$

which, when evaluated gives

$$s_{\bar{C}}^2 = \left[\frac{(bA + a)}{U} s_{\bar{V}}\right]^2 + \left[\frac{V(bA + a)}{U^2} s_{\bar{U}}\right]^2 + \left[\frac{Vb}{U} s_{\bar{A}}\right]^2 + \left[\frac{V}{U} s_{\bar{a}}\right]^2 + \left[\frac{VA}{U} s_{\bar{b}}\right]^2$$

This analysis is important for three reasons:
(1) it allows one to discover which measurement is contributing most to the overall random error and therefore where effort is best put and where it will be wasted
(2) it shows that estimates of precision are usually optimistic because they ignore uncertainty in the standards: direct determination of $s_{\bar{C}}$ by replicate measurements of C is equivalent to assuming that the last two terms are zero
(3) it reveals that there is no such thing as 'the precision of a method'.
The precision will depend on the user and will probably vary from day to day. A numerical example illustrates these points. Using the calibration of Fig. 1.2, suppose V = 50, U = 20, A = 0·24, b = 14·3, a = −0·12. Suppose on 20 replicates of each we measure $s_{\bar{V}} = 0{\cdot}2$, $s_{\bar{U}} = 0{\cdot}2$, $s_{\bar{A}} = 0{\cdot}0002$ and take from § 1.2.3 for n = 6, $s_{\bar{b}} = 0{\cdot}56$, $s_{\bar{a}} = 0{\cdot}20$.

Then $C = \frac{50}{20}(14{\cdot}3 \times 0{\cdot}24 - 0{\cdot}12) = 8{\cdot}3$ concentration units.

$$s_{\bar{C}}^2 = \left[\frac{3{\cdot}31}{20} \times 0{\cdot}2\right]^2 + \left[\frac{50 \times 3{\cdot}31}{400} \times 0{\cdot}2\right]^2 + \left[\frac{715}{20} \times 0{\cdot}002\right]^2 +$$

$$\left[\frac{50}{20} \times 0{\cdot}20\right]^2 + \left[\frac{50 \times 0{\cdot}24}{20} \times 0{\cdot}56\right]^2$$

$= 0{\cdot}00110 + 0{\cdot}00685 + 0{\cdot}00511 + 0{\cdot}25000 + 0{\cdot}11290$
$= 0{\cdot}376$
so $s_{\bar{C}} = 0{\cdot}61$ concentration units.

In this example the error in the slope and intercept of the standard curve contribute (0·250 + 0·113)/0·376 = 66% of the variance, so it is obvious that any attempt to improve precision should concentrate on decreasing the standard error of the calibration, perhaps by using more standards.

If a and b had negligible standard errors then $s_{\bar{C}}$ would be only 0·11. If a particular method has been used routinely one may be tempted to argue that all the standards ever made should be considered and that the error in a and b will therefore be very small. This might be acceptable if the random error of batches were itself random. But this it rarely, if ever, is (§ 1.2.4). 'The true error' of $s_{\bar{C}}$ in this example lies between 0·11 and 0·61. For critical work the conservative 0·61 should be used, but for less demanding routine work 0·11 might be accepted.

This estimation of the relative importance of errors is of course much too complex for everyday work but it is a salutary, and sometimes necessary, proceeding.

In this manual we have in places given estimates of accuracy and precision. It should now be obvious that these are exemplary and not constants for the method.

In general, the larger the number of steps in an analysis which need to be made carefully then the larger will be the overall random error. The worst cases are those where the answer is obtained by subtracting one large number from another only slightly larger one (as in back titrations). The errors are always additive, and if the numbers are large the overall error may exceed the difference. For example suppose

$$p = 142 \text{ ml}, r = 138 \text{ ml, so } p - r = 4 \text{ ml}$$

and suppose the standard errors are $s_{\bar{p}} = 3$ ml, $s_{\bar{r}} = 4$ ml then the combined error (addition of variances) $= \sqrt{9 + 16} = 5$ ml.
So the result is 4 ± 5 ml.
(iv) Sensitivity has many definitions. We define it as the concentration giving a value three times that of the blank, except in atomic absorption flame spectrophotometry where it is twice the background 'noise'. Methods used near their sensitivity limit give unreliable results. Neither multiplication by a large factor during calculation nor quotation of 5 digits in the answer make the result more reliable! The useful range for colorimetric methods is usually from absorbance 0·05 in 5 cm cells up to 0·6 in 0·5 cm cells. The upper limit is roughly 120 times the lower limit if the calibration is a straight line. In other cases the useful range is limited at the point where the calibration line begins to curve markedly. This is rarely a serious problem as the sample can be diluted (but Si determinations are an exception). In volumetric determinations the lower limit is often set by the precision of the burette. With piston microburettes as little as 0·1 ml can be determined with a precision of 1%: that is, of 0·001 ml. The other common limit is set by the extent to which the solution can be diluted without blurring the end point too much. In many cases 0·01 to 0·02 mol l^{-1} is the limit, but 0·001 mol l^{-1} may sometimes be used.

1.2.2 Expression of volumes, amounts and concentrations.

For reasons given in the Preface we have, in this edition, used a compromise between old style units and the SI system. We use old style (SI 'nicksymbols') l and ml. But we have abandoned the 'equivalent' and the 'normal' units. The stoicheiometry* is now shown attached to the chemical symbol. For illustration consider the chemical concentrations:

$$Na^+ \ 1{\cdot}4 \text{ meq } l^{-1} = Na^+ \ 1{\cdot}4 \text{ mM}$$
$$Ca^{2+} \ 1{\cdot}6 \text{ meq } l^{-1} = \tfrac{1}{2}Ca^{2+} \ 1{\cdot}6 \text{ mM} = Ca^{2+} \ 0{\cdot}8 \text{ mM}$$
$$Cr_2O_7^{2-} \text{ (as an oxidising agent) } 0{\cdot}2 \text{ eq } l^{-1} = \tfrac{1}{6}Cr_2O_7^{2-} \ 0{\cdot}2\text{M}$$

and the same solution

$$Cr_2O_7^{2-} \text{ (as a base) } 0{\cdot}6 \text{ eq } l^{-1} = \tfrac{1}{2}Cr_2O_7^{2-} \ 0{\cdot}6 \text{ M}$$

* We use the OED spelling. The OED also gives 'stoechiometry'.

The following table shows the equivalence of some common units.

Equivalence of units

Quantity	Old style	SI system
Volume	l	dm^3
	ml	cm^3
Chemical amount	eq	mol
	(substance from context)	(substance specified e.g. Na^+, $\frac{1}{2}Ca^{2+}$)
Mass concentration	$\mu g\ ml^{-1} = mg\ l^{-1} =$ ppm	$\mu g\ cm^{-3} = mg\ dm^{-3} = g\ m^{-3}$
Chemical concentration	$eq\ l^{-1}$	$mol\ dm^{-3}$
	(substance from context)	(substance specified e.g. Na^+, $\frac{1}{2}Ca^{2+}$)

For the SI concentration unit '$mol\ dm^{-3}$' we often use the abbreviation 'M'. Thus 'mM' means '$mmol\ dm^{-3}$'. 'M' is not standard SI usage but is very convenient.

The results of an analysis are probably best expressed as $mmol\ l^{-1}$ if the compound analysed contributes significantly to the ionic balance (§ 4.10.1). If the mass concentration is needed then $mg\ l^{-1}$ or, to avoid too many leading zeros, $\mu g\ l^{-1}$ are commonly preferred. $1\ mg\ l^{-1} = 1\ \mu g\ ml^{-1} = 1$ part per million (ppm) $= 1\ g\ m^{-3}$.

In some cases the quantities, containers, times and temperatures specified in this manual are not critical, but wanton changes should not be made and the effects of making essential changes should of course be examined. In general we indicate the necessary precision and accuracy by the number of significant digits. The same convention should be followed when giving results. Thus 4 M means 4 ± 1 M whilst 0·100M means $0{\cdot}100 \pm 0{\cdot}001$M. In the first case a measuring cylinder or even a graduated beaker suffices, but in the second a volumetric flask (or apparatus of similar accuracy and precision) is necessary. Similarly a dilution to 0·5 or 1·0 litre suggests the use of a measuring cylinder but dilution to 500 or 1000 ml indicates a volumetric flask. Where a concentration is given as 'percent' then a measuring cylinder will do.

It is expensive to make solutions more accurately or precisely than necessary. It is even more expensive to make them insufficiently so.

1.2.3 Standards

Standards are used to calibrate an instrument or method and to look for interference: that is for systematic errors in a method.

(i) External standards must be used whenever there is no reliable means for calculating a result directly from an experimental measurement. Examples are colorimetric, and flame spectrophotometric methods, and pH measurements. The accuracy of estimation of every sample depends on the standards: it is very unwise to skimp on them even in routine procedures. The practice of using but a single standard or even none at all is highly dangerous. The best that can happen is that a whole batch goes so badly wrong that suspicion is roused and the measurements repeated.

Ideally the standards should be transferred to sample bottles in the field at the same time that samples are collected, but in most cases it is sufficient to introduce standards just before the analytic reaction proper.

It is usual to use standards spaced at equal concentration intervals, for example 0,2,4,6,8,10. But if most of the samples are expected to fall in the range 0 to 2 then it may be better to put more standards into that range, for example 0, 0·5,1,2,4,8. One should be suspicious of any result which falls between the blank and the lowest concentration because there may be unsuspected effects at very low concentration. Equally one should suspect results which require extrapolation beyond the uppermost standard. It may be helpful to duplicate points: 0,0, 4,4, 8,8 because this gives immediate indication of precision (reproducibility). More usually however some form of curve fitting—subjective or objective—is used and the spaced points are then more helpful. An estimate of precision can still be had from such an arrangement.

In the simplest, and fortunately the commonest, case Beer's law is obeyed and there is a linear relation between absorbance and concentration. The equation of this is

$$y = bx + a \qquad (1.1)$$

where y = concentration (dependent, because in use the absorbance is known)
x = absorbance
b = a 'constant', the slope, with units y/x
a = a 'constant', the concentration at zero absorbance, units of y.

If the machine has been *set* to zero with a 'blank' (§ 1.4.2) then the value of 'a' should be zero or very close to it. For routine work it may suffice to use a ruler and draw the line subjectively. Even then it is well worth the effort of calculating the mean of all 'x' and the mean of all 'y', then making sure the line goes through that point. Values of concentration corresponding to measured absorbance can then be read from the graph but this reading process is liable to result in mistakes because one must read across from the y axis to the line and then turn at right angles to read down to the x axis.

If the intercept 'a' is nearly zero, or if a simple calculator is to hand it is quicker and more reliable to measure 'b' (and 'a') and calculate the concentrations from equation 1.1.

If the standard points are more scattered, but still apparently linear it may be worth while calculating the linear regression (using sums of squares of deviations of y as a criterion) to get 'b' and 'a'.

In what follows the symbol 'Σ' means 'sum for all items'.
Using figures from the numeric example which follows this section:

Σy means $(0 + 3 + 6 + \ldots\ldots\ldots\ldots + 15)$
Σy^2 means $(0^2 + 3^2 + 6^2 + \ldots\ldots\ldots\ldots + 15^2)$
Σxy means $(0{\cdot}004 \times 0 + 0{\cdot}201 \times 3 + 0{\cdot}449 \times 6 + \ldots\ldots\ldots + 1{\cdot}005 \times 15)$

Note that $(\Sigma y)^2$ means 45^2 and is not the same as Σy^2.

The equations which follow may look forbidding at first sight but they have remarkable symmetry and once understood are easily memorised.

For convenience define

$C_{xy} = \Sigma xy - \Sigma x \Sigma y/n$
$C_{xx} = \Sigma xx - \Sigma x \Sigma x/n$ but calculate as $\Sigma x^2 - (\Sigma x)^2/n$
$C_{yy} = \Sigma yy - \Sigma y \Sigma y/n$ but calculate as $\Sigma y^2 - (\Sigma y)^2/n$

Then

$$b = C_{xy}/C_{xx}$$
$$a = \bar{y} - b\bar{x}$$

where $\bar{y} = \Sigma y/n$ = mean value of y

$$\bar{x} = \Sigma x/n$$

It is easy to remember that the slope 'b', which is measured by (y axis length)/(x axis length), must be C_{xy}/C_{xx} because this gives the correct dimensions of y/x.

The intercept 'a' on the y axis is got using the fact that the line of best fit must go through the point $(\bar{x},\bar{y})$ so that

$$\bar{y} = b\bar{x} + a$$

If the standard errors of a and b are needed we must also calculate:

$$s_y^2 = \frac{C_{yy}}{(n-1)} \cdot (1-r^2)$$

where r is the correlation coefficient and

$$r^2 = \frac{C_{xy}{}^2}{C_{xx}.C_{yy}}$$

Then

$$s_a = \sqrt{s_y^2/n} \quad = \text{standard error of a}$$
$$s_b = \sqrt{s_y^2/C_{xx}} = \text{standard error of b}$$

The utility of these last statistics is explained in § 1.2.1.

A numeric example follows. The data are plotted in Fig. 1.2.

Absorbance x	0·004	0·201	0·449	0·641	0·895	1·005
Concentration y	0	3	6	9	12	15

$n = 6$		
$\Sigma x = 3{\cdot}2$	$\Sigma y = 45$	
$\Sigma x^2 = 2{\cdot}464$	$[\Sigma y^2 = 495]$	$\Sigma xy = 34{\cdot}881$
$\bar{x} = 0{\cdot}533$	$\bar{y} = 7{\cdot}5$	
$C_{xx} = 0{\cdot}763$	$[C_{yy} = 157{\cdot}5]$	$C_{xy} = 10{\cdot}919$
$b = 14{\cdot}32$	$a = -0{\cdot}12$	$[r = 0{\cdot}996 \quad s_y{}^2 = 0{\cdot}235]$
$[s_b = 0{\cdot}56$	$s_a = 0{\cdot}198]$	

With a hand calculator the calculation of a and b takes about 3 minutes. The extra items in brackets take a little longer.

In a few methods the calibration line is always curved: we have avoided such methods as far as possible. The occurrence of a curved calibration line with a method which normally gives a straight line is probably the result of a mistake. For curved lines a different equation is necessary and calculation is more lengthy. Those with access to a computer may choose to fit a polynomial but the simple graphical plot will usually be cheapest and quickest.

The standard line should never be extrapolated, that is it should never be used outside the range for which it was measured, because it may well cease to be linear. Such dangers are more easily overlooked if the results are calculated than if they are read from the graph. The dangers of extrapolation are even greater for calibration lines which are conspicuously non-linear within the measured range.

The zero concentration standard is often called the 'blank'. If treated as just one of the many concentrations it has no special significance, but in colorimetric work it is often used in a special way (§ 1.4.2) so that all other measurements, standards and samples, depend on it. The need for replication if this is done should hardly need emphasis.

(ii) The method of 'standard addition' is used to seek for and measure 'interference', which is one type of systematic error. The method is sometimes called 'internal standards' but this term has another and particular meaning. The term 'spiking with standards' is also used.

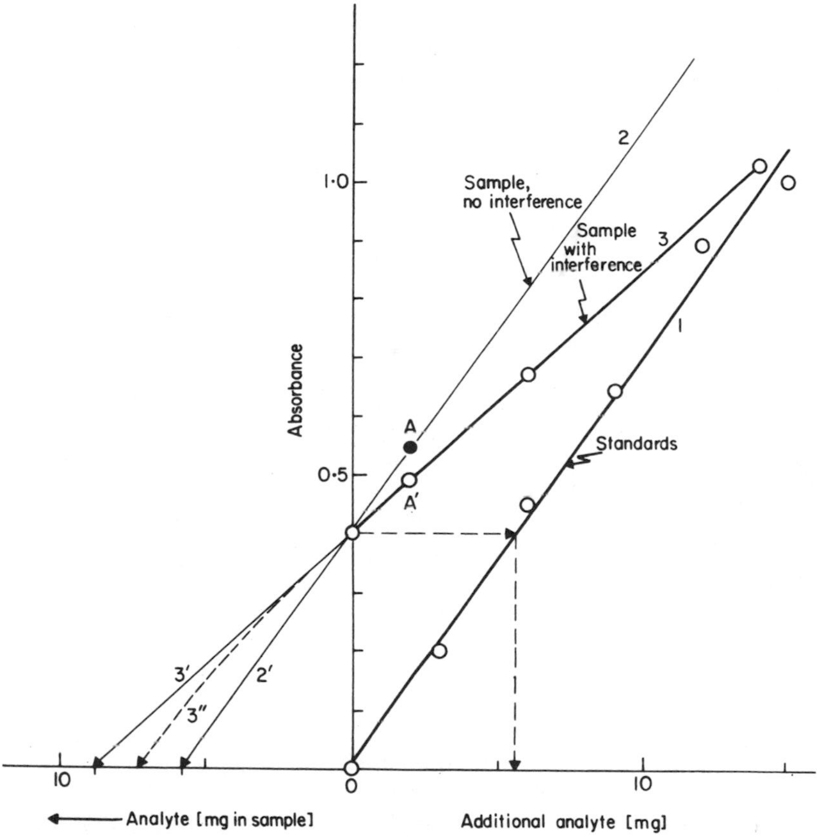

Figure 1.2. Standards. Note that numbers to the left of zero run in the opposite direction from those to the right.

Line 1: absorbance of known standards following Beer's law.

Line 2: predicted absorbance, if there is no interference, when standards are added to a sample with absorbance 0·40. In particular A is the predicted absorbance if 2 mg of analyte are added. The amount in the sample, assuming no interference, may be got either by the usual process of reading off the value on the standard line 1, or by extrapolating back along 2′ to the x axis. It will be obvious (from simple geometry) that the intercept to the left of zero is the same length as that to the right.

Line 3: actual absorbance. In particular A′ is the absorbance when 2 mg of analyte are added. The increase in absorbance is only 65% of that expected so there is interference. The amount of analyte in the sample is estimated using 3 different additions and extrapolating along 3′.

Line 3″: a possible form of behaviour which would invalidate the estimate.

Suppose for example that we use a colorimetric method. Standards are made, using the analyte and H_2O. The absorbance of these, and of the sample, is measured. The sample absorbance is 0·400. From the standard curve (Fig. 1.2) we calculate that if there is no interference the sample contains 5·8 mg and predict that the addition of a further 2 mg of analyte to the sample should increase absorbance to 0·540. When this is done the measured absorbance is 0·490—points A and A′ in Fig. 1.2. The addition has had only 65% of the effect expected. If the calibration is linear we might conclude that the original amount is not 5·8 mg but 5·8 × 100/65 = 8·9 mg.

If there is interference at all however it is unwise to assume it is linear. Add different amounts of analyte to at least 3 sub-samples. The largest addition should be at least the amount which is thought to be in the sub-sample. Then measure absorbance. This amounts to constructing a calibration curve for each sample individually. The estimate of analyte amount is got by extrapolation to zero absorbance (Fig 1.2).

The estimate depends on extrapolation, which is always dangerous, and particularly so if the shape is non-linear. Even if it is linear in the measured range of addition there is no guarantee that it is so at lower levels—pecked line in Fig. 1.2.

Some forms of chemical interference are not revealed by the method of standard addition. The interfering substance may for example behave chemically in the analysis in the same way that the analyte does. Iodide may be confounded with chloride, and arsenate with phosphate.

It is preferable therefore to try to remove the cause of interference. General methods are given in § 1.4.2 (colorimetry) and § 1.4.3 (flame spectrophotometry). Specific methods are given in the appropriate places.

It will probably be obvious that the quantity to be plotted against absorbance on calibration graphs depends on the circumstances. If all standards and samples are made with the same volume and made up to the same volume then mass or chemical amount (perhaps μg or μmol) are probably the most generally useful. But in some cases concentration in the volume taken, or (as in the example of random error combination) concentration in the final volume, may be preferable.

1.2.4 Laboratory practice

There are a number of points which are rarely mentioned in analytical chemistry texts but which contribute to efficient working.

First, it often happens that when a method is tried for the first time, or when it is restarted after a break of some weeks, then precision (reproducibility) and accuracy are unsatisfactory (Young, 1949). Simply repeating the procedure, with no deliberate changes, will often restore the earlier reliability. But it is wise to assume that the first batch of measurements will not be satisfactory and to avoid including samples which cannot be repeated.

Secondly, it should be constant practice when a new standard solution is made to check it against the old one, and sufficient of the old one should be kept for this purpose. This is necessary even if commercial standard solutions are used.

Thirdly, the analyst should constantly be seeking to match the precision and accuracy of his analyses to the needs of his problem and the limitations of the sampling programme. It is usually a waste of money to strive for accuracy ± 1% if the samples are unrepresentative or have a standard deviation of 10%, or if the problem involves the comparison of two samples which differ in concentration by a factor of 10 (but see § 1.7).

Even within a single procedure there is scope for judgement. The standard solution may have to be made with accuracy ± 0·1%, the HCl used to make the solution acid may need to be of accuracy no better than ± 10%, but it may have to be dispensed in amounts which have precision (reproducibility) ± 0·5%. The acid should then be made up rapidly in a measuring cylinder and dispensed with an automatic (syringe) dispenser. But if the laboratory needs stocks of both 0·1M and 0·100M HCl then using the more accurate 0·100M solution for all purposes may cost less and cause fewer mistakes to be made.

We have not indicated in the methods which operations must be of high accuracy and which merely reproducible because we believe that the interested worker will soon discover on which operations time may be saved. For those who insist on using this manual simply as a recipe book it is better to waste some time than to get invalid results.

The limnochemist interested in reducing costs may care to consider converting operations from a volumetric to a mass basis. Volumetric glassware is now very expensive whilst ungraduated glassware has not increased in price by the same proportion. Most laboratories now have top loading (= open pan = top pan) balances which are reliable and more accurate than volumetric glassware. To make solutions up to a given mass is at least as quick as to make them to a given volume. And it is possible to measure an existing mass much more quickly and accurately than it is to measure its volume or to make it up to an accurately measured volume.

1.2.5 Cleaning of glassware

The classic glassware cleaner is chromic acid.

In 1 litre of H_2O dissolve 100 g of $K_2Cr_2O_7$. Add slowly, with great care and whilst stirring, 1 litre of concentrated H_2SO_4. Use hot, or allow the glassware to soak overnight. This mixture is hygroscopic and caustic so containers should be covered.

Chromic acid is dangerous and for many purposes one of the safer modern proprietary glassware cleaners may be suitable. They are complex mixtures so it is essential to make sure that they do not interfere in analyses. Ordinary detergents are less effective and more likely to cause interference. They may have disastrous effects on phosphate analyses.

1.3 REAGENTS AND CHEMICALS

1.3.1 Reagent grades

Many of the reagents specified in the methods must be of high purity. Different makers use different names for this grade. 'Proanalyse', 'Analytical Grade', 'Analar' and 'Baker Analysed' are examples. In this manual the abbreviation AR (for Analytical Reagent) is used. If the chemical is to be used as a standard it is usually necessary to dry it in an oven at 105°C for one hour to remove traces of water. The fact that the concentration of impurities is known allows one to calculate how much of the compound to be analysed will be introduced.

The AR grade of purity (and expense) is not always necessary. The cheaper 'Laboratory Grade' or 'Purissimus' may then be used.

1.3.2 Water

The most important reagent is H_2O. In this manual H_2O means distilled water collected in a container which does not release substantial amounts of substances into the water. Flexible plastic tube very often does release organic 'plasticisers', more rapidly if the water is hot. It may be necessary to redistill the water, particularly for NH_3 determination. If a second distillation is made then the purpose for which this water is to be used should be considered. To remove the last traces of NH_3 for example the second distillation is made after adding H_2SO_4 to the first distillate.

The use of ion exchange resins alone has often been reported but many of the organic substances in tapwater are not removed completely. Indeed some ion exchange resins may release organic substances to the water.

Ion exchangers may occasionally be useful before or after the distillation stage, but not all commercial equipment is satisfactory for limnochemical use.

The quality of distillate may be checked by measuring the electrical conductivity which should be less than 5 $\mu S\ cm^{-1}$. If the conductivity increases in the course of time then the distillation apparatus probably needs cleaning. Low conductivity does not mean the water is free of organic compounds however.

1.3.3 Making up solutions

Where two liquids are to be mixed it may be best to add the larger volume to the smaller one, because most of the mixing is effected during the addition of the two liquids. But with H_2SO_4 the heat of dilution is so great that the solution may boil violently and cause drops of liquid to be thrown out. H_2SO_4 should therefore be diluted by adding the acid slowly, carefully and with stirring, to water.

Where a solid and liquid are to be mixed it is usually best to add the solid to the liquid, stirring the while because the solid then disperses. Readily soluble solids may otherwise form a 'cake' with the first drops of liquid, and then be difficult to disperse. It may occasionally be useful to make use of the heat of solution to aid dissolution by adding the solid to only a small part of the liquid.

Mixing must always be thorough and careful.

1.3.4 Stock solutions

The unspecific instruction 'dilute' means 'dilute with H_2O'.
It is useful to have a stock of the following solutions:

HCl, 12 M
'Concentrated', 37%, specific gravity 1·18

HCl, 4 M
1 litre 12 M HCl diluted to 3 litre

HCl, 1 M
0·25 litre 4 M HCl diluted to 1·0 litre

HCl, 0·100 M
25 ml 4 M HCl diluted to 1 litre. Standardise with 0·100 M NaOH

NH_2SO_3H (sulphamic acid), 0·100 M
Dissolve 9·71 g of NH_2SO_3H (AR) in H_2O and dilute to 1000 ml. This acid may usually be used in place of HCl in acid-base titrations. It needs no standardising

H_2SO_4, 18 M
'Concentrated', 96–98%, specific gravity 1·84

H_2SO_4 (1 + 1), about 10 M
To 0·5 litre H_2O add cautiously, slowly and with stirring 0·5 litre 18 M H_2SO_4
Warning: never add H_2O to concentrated H_2SO_4

H_2SO_4, about 2 M = ½ H_2SO_4, about 4 M

To 0·8 litre H_2O add 0·2 litre 9 M H_2SO_4

NaOH, 10 M

In 0·5 litre H_2O dissolve 400g NaOH. Dilute to 1·0 litre

NaOH, 0·100 M

10 ml 10 M NaOH diluted to 1·00 litre
Standardise with oxalic acid

Oxalic Acid ½$(COOH)_2$, 0·100 M

Small crystals of the hydrate $(COOH)_2.2H_2O$ are stable between 5% and 95% relative humidity.

In about 0·1 litre H_2O dissolve 6·300g $(COOH)_2.2H_2O$ (AR). Dilute to 1000 ml and store in a refrigerator.

This solution is reliable for months and is the ideal primary standard for both acid-base and oxidimetric titrations

Dichromate, ⅙$(K_2Cr_2O_7)$ 0·100 M

Dry $K_2Cr_2O_7$ (AR) for 2 hours at 105°C. In about 0·1 litre H_2O dissolve 4·903g of $K_2Cr_2O_7$. Dilute to 1000 ml.

Na_2EDTA, 0·100 M = ½ Na_2EDTA, 0·200 M

Dry Na_2EDTA (AR) at 80°C. In about 500 ml H_2O dissolve 37·22g of Na_2EDTA. Dilute to 1000 ml.

Standardised solutions may be bought. Their use saves a lot of time.

1.4 NOTES ABOUT PARTICULAR TECHNIQUES

Some techniques are used in many methods. It is convenient therefore to discuss them here

1.4.1 Volumetric titrations

For level I methods a calibrated pipette or even a simple plastic syringe will suffice, but for level II it is essential to use a reliable piston microburette. Work is expedited if the burette is attached to a reservoir of titrant and has an automatic zeroing device. An accuracy of ± 0·001 ml is necessary with a delivered volume of 2 ml.

The concentration of compounds to be determined is sometimes two or three orders of magnitude less than is common in classical analytical chemistry and the addition of an indicator must then be avoided. Devices for detecting electrometric end points are thus needed. A conductance meter and pH meter of normal laboratory quality are sufficient. For pH determination an accuracy of ± 0·02 units suffices and allows one to make acid-base and potentiometric titrations (including 'dead stop titrations' as described in § 9.13).

In many volumetric determinations the titrant is itself the primary standard and the reaction is stoicheiometric so it is not necessary to use standard solutions of the substance being determined. But it may nevertheless be useful to use standards as a check or to evalute a new method.

1.4.2 Colorimetric determinations

(*i*) *Colorimeters and spectrophotometers*

At level I colours are usually matched directly by eye, but at level II the match is indirect: standards and samples are measured individually on an absolute scale and then the sample reading is compared with a calibration graph drawn from the standard readings. Either a colorimeter or spectrophotometer may be used. Both have a stable light source from which the required colour is selected and passed through the sample to a detector. In the colorimeter a coloured filter absorbs most of the light and transmits only a narrow waveband which includes the required colour. The filter may be changed and is chosen to suit the analysis. In the spectrophotometer the light beam is dispersed by a prism or grating into a rainbow coloured strip from which the required spectral range is selected through a narrow slit by mechanically rotating the prism or grating, thus sweeping the rainbow across the slit. The light beam may be modulated or 'chopped' to reduce interference and drift. More expensive instruments commonly have a 'double beam.'

The spectrophotometer allows one to select a narrower spectral range than the colorimeter so that interferences are reduced. But it needs a more sensitive detector and is mechanically more demanding so is more expensive than a colorimeter. The mechanical linkage of a spectrophotometer can drift out of adjustment so the wavelength calibration should be checked, for example with a didymium glass standard, at least once a month. If necessary the calibration should be adjusted. The colorimeter depends on the physical properties of the filter and does not need adjustment. Operators are less likely to select the wrong filter than they are to set the wrong wavelength: 426 nm instead of 462 nm for example.

The wavelength at which absorbance is measured is in the range 380 to 800 nm for most methods and is usually that of a peak (or the only peak) in the visible absorption spectrum. At this wavelength the method is most sensitive and is little affected by small changes in the wavelength setting of a spectrophotometer. There is usually no reason however why some other wavelength should not be used, and this may be advisable if the water contains some natural coloured substance which, after addition of all reagents except that producing the analytical colour, has appreciable absorbance at the wavelength normally used. When using a spectrophotometer one should choose a flat part of the absorption spectrum, so that a small change in wavelength has little effect. If for some reason (for example in Si determination) a sloping part of the absorption curve must be used then it is preferable to use a colorimeter because the wavelength setting cannot drift.

If a spectrophotometer or colorimeter has a meter which responds linearly to absorption of light and if the errors of all sorts in the reading are independent of the reading, then it can be shown (Kolthoff and Sandell, 1952, p 631) that the relative error in absorbance is least at 0·43 and becomes markedly larger outside the range 0·05 to 0·6. These assumptions are not strictly correct for many modern spectrophotometers and colorimeters, but it is still a useful guide to try to arrange that sample absorbance falls within the range 0·05 to 0·6. This is most easily done by choosing the path length of the cell. A set of 5 mm, 10 mm and 50 mm cells should normally suffice.

The statement 'measure the absorbance in suitable cells', which appears in many of the methods, is intended to mean that the cells are chosen so that the absorbance falls between 0·05 and 0·6.

When two or more cells are used two types of error may be introduced. The first results from unequal absorption of light by the glass of the cells. This causes a constant difference in absorption between the cells. The second error results from small differences in the dimensions of the cells which affects the path length. This causes a constant proportional difference in absorbance. Both errors are usually negligible. If the first is so then the second may be corrected by measuring the absorbance of the same test solution in all cells and calculating from these measurements a cell correction multiplier (value nearly 1·0) which is applied to the measured absorbance.

(*ii*) *'Blank' value for a colorimetric determination*

The reagent blank is distilled water treated exactly as are the samples. It is essential to make one reagent blank and strongly advisable to make at least two with each analysis batch, because the information derived from the 'blank' may be used in the calculation of absorbance of all standards and samples.

If the absorbance of the reagent blank is normally small then it may be convenient to avoid calculations by first checking that the absorbance of both blanks *is* small and differs by less than 0·002, and then, with one of the blanks in position, to set zero absorbance on the machine. If these conditions are not met then it is preferable to use distilled water as the blank. The absorbance for zero concentration may then be estimated from the standard curve and subtracted from all other measurements (§ 1.2.3).

If the water sample itself, before treatment, absorbs more strongly than does distilled water in the spectral region to be used in analysis then some inaccuracy will result. In many cases this error is trivial: the fact that it can be detected should not in itself be a cause for worry. If however the error is not trivial, as for example in many acid peaty waters, or if high accuracy is needed, then there are three possible remedies.

First it may be possible to shift to a different wavelength at which the natural colour is negligible but the analytical one is still marked.

Secondly it may be possible to destroy the natural colour without affecting the analysis. Heating with $K_2S_2O_8$ may be suitable.

Thirdly some form of correction may be made to the absorbance measured after colour development. There is unfortunately no entirely satisfactory method of doing this. The simplest is to measure the absorbance of the untreated water and to subtract this value from the absorbance after colour development (after allowing for dilution of the sample by reagents). In many cases however the added reagents alter the absorbance of the natural coloured compounds. Change of pH in particular often causes change in absorbance. A better correction may therefore be obtained by adding all reagents which affect the pH markedly before measuring the 'natural' absorbance. In difficult cases 'standard addition' (§ 1.2.1) may be necessary.

The nearest to the ideal is found in NO_2^- analysis (§ 5.4.1) where all reagents may be added, but colour develops only if they are added in the correct order.

1.4.3 Flame emission and atomic absorption determinations

Volumetric and colorimetric techniques have a long history of development. Flame emission and atomic absorption methods have become practicable and commonplace only in the last 20 years. Many of the standard books on analytical chemistry do not mention them. Atomic absorption methods in particular are distinguished by the fact that a relatively expensive machine is needed but once obtained that machine is capable of estimating the concentration of over 60 metallic and semi-metallic elements in complex mixtures, some to 1 part in 10^9, relatively free of interference, needing only 1 to 10 ml of solution and often needing no measurement of solution volumes. Because of these features such machines are now fairly common but they *are* complex machines and the untutored user should be specially wary.

Principles and instruments

In both methods the sample solution is sucked at constant rate into a flame of carefully controlled and stable character. The ground state atoms absorb thermal energy and some electrons are raised to a higher energy level. A proportion fall back to a lower state releasing energy as light of characteristic wavelengths, for example the yellow of sodium. In flame emission work the flux of emitted light is measured, and can be related to the concentration in the solution. The proportion of atoms emitting light is small: perhaps 1 %, but the ground state atoms have a highly specific and very sharp absorption spectrum. If light of one of these wavelengths is generated outside and passed through the flame then absorbance follows the Beer and Lambert laws: it is proportional to path length and to concentration (in the flame). This is the method of atomic absorption. It is similar to spectrophotometry, with the flame replacing the cuvette. In flame emission work the light is produced within the flame: in atomic absorption work it is produced by separate apparatus outside the flame.

Some ions (Na^+, K^+, Li^+, Ca^{2+}) have intense emission lines in a cool air-propane flame. It is practicable to use a simple photocell as light detector and coloured filters to select the appropriate wavelength. The relatively simple flame emission colorimeter is the result. In arrangement it is similar to the colorimeter, but interest lies in the character of the light source, rather than in what happens to the light. Where greater sensitivity and selectivity are needed the filter and photocell may be replaced by a prism (or grating) and photomultiplier tube. The result is the flame emission spectrophotometer.

In atomic absorption instruments light of the specific wavelength of interest is produced by a hollow cathode lamp. This has a cathode made of the element to be analysed. It emits light which is mostly at just those wavelengths which are needed. In a spectrophotometer the equivalent is the broad spectrum lamp 'masked' by a prism (or grating).

It is now possible to get hollow cathode lamps containing several elements. These are more expensive than single element lamps but much less so than separate lamps for each element. They may be a sensible choice for those elements which are infrequently examined.

The light then passes through the flame where it is attenuated. But the flame also emits unwanted light at other wavelengths. This unwanted light is diverted from the detector by a prism (or grating) which has no analogue in the spectrophotometer. The light detector is a photomultiplier tube. This complete instrument is the atomic absorption flame spectrophotometer (AAFS). It is perhaps ten times as expensive as the flame emission colorimeter, but of higher stability and accuracy and has much greater flexibility. It can usually be used for flame emission work too, omitting the hollow cathode lamp, perhaps making small changes to the instrument, and measuring emittance rather than absorbance.

In all but the simplest atomic absorption spectrophotometers it is possible to use a variety of types of flame. In order of increasing temperature the commonest are: air-propane, air-acetylene, nitrous oxide-acetylene. By adjusting the proportions of oxidant and fuel gas it is possible to work in reducing or oxidising conditions.

There are several patterns of burner. Most of those on commercial instruments are of the 'premix' type: the fuel and oxidant gases are mixed with solution in a chamber and then pass at high velocity out of the burner slot into the flame. The optical path (along the length of the flame) is long, the burner is not easily blocked droplet size is uniform and the burner is quiet.

The solution to be analysed is sucked into the machine using the Pitot tube effect. It is hurled against a glass bead which breaks it up into a fine spray. The larger droplets condense on baffles and run to waste. Only the mist reaches the flame. It will be obvious that the exact position of the 'nebuliser bead' is critical to the performance of the machine, and in most it can and should be adjusted by the user.

The gas mixture in the premix chamber is potentially explosive and a modicum of care is needed to avoid unpleasant incidents:

(i) The waste solution drain contains a U shaped tube which must be kept full of liquid. This liquid seal should be checked before lighting the flame.

(ii) Any occlusion of the burner slot may increase pressure in the premix chamber and blow the liquid out of the seal. The commonest cause of this is salt deposition in the burner. Alternating samples and H_2O will usually prevent this, but if salt concentrations are high (5 to 10% w/v) then a three slot Boling type burner may be preferable if available.

(iii) The limnologist will probably have no choice of burner type: this need occasion no alarm. But it is essential to use the correct *size* of burner slot for the gas mixture selected. The nitrous oxide-acetylene flame should not be used with an air-acetylene burner. Follow the maker's instructions.

Some machines have an automatic gas control and flame lighting attachment. This minimises most of these dangers, but no device is completely proof against ignorance.

In general the optical and electronic parts of AAFS instruments can be made to high and constant stability. The overall performance is usually limited by the characteristics of the flame, and these are the responsibility, and should be within the control of, the user.

For many elements the calibration curve is nearly linear over a wide range of concentration. For some it is not, and all become non-linear at high concentration. Some machines allow one to apply a standard 'correction' to straighten the curved part (Fig. 1.3). For rough work this is adequate, but the user should still check with standards that the automatic correction is sufficiently accurate. In many cases the correction is only necessary when absorbance is greater than 0·6. As with spectrophotometric work one should try to work within the range of absorbance 0·05 to 0·6. If the solution is so concentrated as to give absorbance greater than this then it may be diluted, but it is often quicker to move to another part of the absorption spectrum of lower sensitivity or, in some cases, to rotate the burner so that the light path through it becomes shorter. Of course one must then recalibrate and beware of interference because a much larger proportion of the light path lies in non-ideal (cooler or more oxidising) parts of the flame.

Many machines also allow one to calibrate directly in chosen units of concentration by applying electronic multiplication ('scale expansion') to the absorbance reading. If this aid is used it is essential to check that the absorbance reading is roughly correct before calibrating in concentration units. Suppose that this is not done and, because one of the many adjustments has been incorrectly made, the absorbance for a given standard is 0·01 instead of the 0·5 which it should be. Tiny fluctuations in flame character may produce random movements of 0·001 in absorbance. This is 10% of the actual absorbance instead of the 0·2% which it should be. In concentration mode the machine will appear to be (and indeed is) very unstable. But the fault lies in the user not in the machine.

The limnochemist will often be working near the lower limit of detection. Short term fluctuations in absorbance are much greater than in spectrophotometry

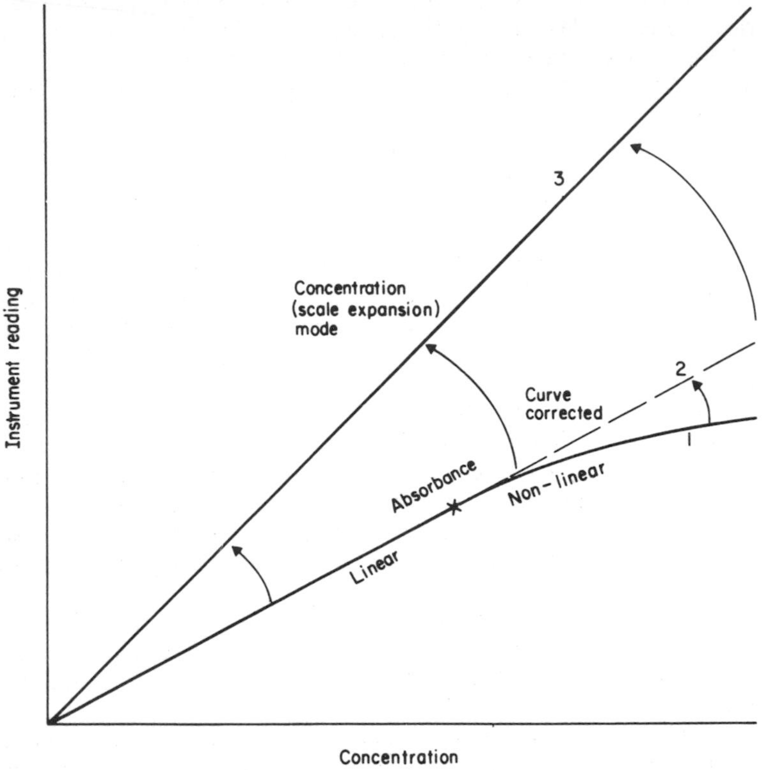

Figure 1.3. Effects of successive use of 'curve correction' (line 2) and 'concentration mode' or 'scale expansion' (line 3) on a non-linear calibration line. Devices to make these changes are commonly found on atomic absorption flame spectrophotometers and on the more expensive modern spectrophotometers.

and it is often difficult to determine the average reading of a galvanometer needle unless this is heavily damped (by a capacitor across the terminals). If it is so damped the needle may take 30 seconds to get within 0·001 of the true value. One solution is to record the reading on a chart plotter, and then to take a visual average. But the rare large excursions from the mean look misleadingly important. Some machines have digital sampling devices which take an average of many (commonly 10 to 100) readings during 1 to 10 seconds. These devices can save a lot of time and remove the subjective element from the reading process.

Starting up an atomic absorption spectrophotometer is much more complex than starting up a spectrophotometer. Commonly the following adjustments must be made:

Hollow cathode lamp orientation and current, wavelength selection and adjustment, slit width, choice of gases, burner and pressures, adjustment of burner height, orientation and lateral position, adjustment of solution flow rate and nebuliser bead position.

Failure at any of these points may result in no readings at all. Perhaps worse is to get readings with large systematic errors in them.

One disadvantage of flame spectroscopy is that the analyte is swept through the flame very rapidly. A development of the method known as 'carbon arc' or 'graphite furnace' spectroscopy replaces the flame by a tiny graphite furnace. The solution or solid is put inside, then heated to white heat. The atoms stay in position for an appreciable time. About 5 mg of solid or solution are sufficient and the sensitivity is perhaps 100 times greater. The technique is more difficult than flame spectroscopy however.

Other types of 'flameless' methods may be used in some cases: Hg and As for instance.

Interferences

Interferences may be detected by the method of standard addition described in § 1.2.

AAFS is relatively free of interference, particularly the 'self colour' type which is so troublesome in colorimetry. But there are five types of interference of which the analyst should be aware. Flame emission methods suffer more than atomic absorption ones.

(*i*) *Aspiration rate*

Anything which changes the rate of aspiration of solution will alter the concentration of atoms in the flame. The commonest intermittent causes are a small obstruction in the aspiration tube or the tube being so close to the wall of the sample container that flow of liquid is impeded. Similar interference may occur if salt deposits occlude the burner. Changing to a different solvent with altered viscosity, surface tension, density and perhaps combustibility will also cause interference.

(*ii*) *Formation of insoluble compounds*

Insoluble chlorides and sulphates are a common cause of such interference, but if the water sample has been filtered and no additions have been made such interference should not be met. If additions must be made it is preferable to use nitrates. Some elements form oxides in the flame. The hotter and more reducing nitrous oxide-acetylene flame will usually minimise interference from both these sources. If it does not or if it cannot be used for some other reason then it may be possible to prevent the interference by adding a competitor. For example interference by phosphate in the determination of calcium may be prevented by addition of $Sr(NO_3)_2$ or $LaCl_3$.

It is possible, though tedious, to extract analyte or interferent, though extracting the analyte could be used to increase concentration too.

(*iii*) *Ionisation of the element*

This is a problem with alkali and alkaline earth elements (including Na, K, Ca, Mg) which ionise appreciably in flames hotter than air-propane. Absorbance is reduced in proportion to the number of atoms ionised. Such interference may be reduced by using an air-propane flame or by adding a large excess—2000 to 5000 ppm*—of a more easily ionised element. Cs is particularly useful.

(*iv*) *Unspecific non-atomic absorption* (*as for example by small solid particles in the solution*)

This absorption is characterised by having a broad spectrum rather than sharp lines. It is sometimes possible to find a nearby wavelength at which the lamp emits but at which the atoms of the analyte in the flame do not absorb. This wave-

* AAFS is often used for measuring concentrations of trace elements, so the unit ppm (parts per million) is standard. Concentrations of interference suppressors are often given in the same units. In a solution 1 ppm = 1 mg l^{-1}

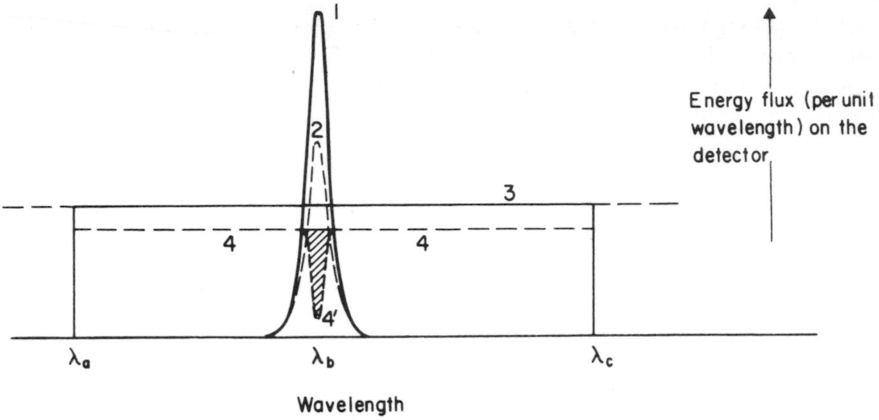

Figure 1.4. Principle of correction for non-atomic ('molecular') absorption in AAFS. The analyte lamp emits the peak 1 centred about wavelength λ_b. When the sample is introduced the smaller peak 2 results giving absorbance 0·22. The hydrogen continuum lamp emits the spectrum 3 which is cut off at λ_a and λ_c by the spectrometer slit. When the sample is introduced the spectrum 4,4′ results. This has absorbance 0·10 and is mostly a result of molecular absorption. The shaded trough 4′ absorbed by the analyte is less than 5% of the whole and is ignored. The corrected analyte absorbance is then 0·22 − 0·10 = 0·12. The correction is usually much smaller than this, commonly 0·002 or less.

length may then be used to measure non-atomic absorption. Of much more general value however is the hydrogen continuum lamp. This emits a broad spectrum of radiation. Figure 1.4 shows how it can be used to correct non-atomic absorption. For all but the most accurate work it is usual to ignore the small absorption by the analyte in the hydrogen continuum and to subtract the absorbance in the hydrogen continuum directly from that with the analyte lamp in position.

Some machines make this correction automatically.

(*v*) *Overlapping spectra*

These may be difficult to avoid in flame emission work. The unwanted emission may come from other elements in the solution or from the flame itself. The nitrous oxide-acetylene flame is particularly troublesome. There is little to be done but move to another part of the spectrum. This interference can almost always be avoided in atomic absorption work, and is indeed one of the main advantages of the technique.

The limnochemist should be aware of but not discouraged by the forbidding list of possible sources of interference. In many cases there will be none.

1.4.4. Gas chromatography

Any volatile substance may be determined by gas chromatography. In this manual are methods for N_2, CO_2, CH_4 and O_2. It can also be used for many organic substances, including chlorinated hydrocarbons and lipids. Such analyses are at Level III.

In principle the volatile substance is removed from the water sample in a carrier gas. The mixture is dried and injected into the end of a long narrow column containing the stationary phase—a porous solid or a non volatile liquid supported on an inert solid. Carrier gas passes through the column. The sample gases become partitioned between moving and stationary phases to differing degrees and therefore travel along the column at differing rates. The concentration of gas is measured as it emerges from the end of the column.

As little as 10 ml of water sample is needed and several compounds may be measured on each sample.

1.4.5 Specific ion electrodes.

Ions in solution may be caused to produce an electrical potential in the manner described in § 9.9. The electrode is made specific by restricting the access of all but the one ion of interest by using a membrane or a specific solid state reaction.

The earliest of these electrodes, the 'glass electrode' used for measuring pH, is of great value. The potential difference across the thin glass bulb is directly proportional to the difference in pH across the bulb.

Ion selective electrodes have several advantages:

(i) Ease of operation: similar to the measurement of pH
(ii) Direct measurement of ion activity (rather than of concentration)
(iii) Rapid response
(iv) Wide range of logarithmic response from as low as 1·0 or even 0·1 μM over five or more orders of magnitude. The potential change is 50 to 60 mV for a tenfold change in concentration so measurements cannot be highly precise.

There are two difficulties associated with these electrodes. First their useful concentration range is sometimes above that needed by the limnochemist. The NH_4^+ electrode is an example. Secondly they are by no means so specific as the name suggests. For example Cl^- electrodes are affected to some extent by the concentration of OH^-, Br^-, I^-, and S^{2-}. We have also found that one type at least is sensitive to humic acids. The K^+ electrode is sensitive to NH_4^+.

They may have use in special circumstances. For example the NH_4^+ electrode may be used to control water quality in a fish tank and the Cl^- electrode to monitor chlorinity in a tidal area. The great advantage of such electrodes is that the signal from them may be used to initiate control actions. Whitfield (1971) gives details.

Some of these characteristics are also found in 'electrodes' for oxygen concentration measurement.

1.5 SAMPLING

There are two aspects to the problem of sampling. The first is when and where to sample. The second is how to transfer the samples with minimal chemical change to the place where analyses are made.

1.5.1 Place and time of sampling

No manual can hope to give comprehensive directions about where samples should be taken if only because this is much affected by the purpose of the work. A modest acquaintance with simple statistical theory is essential, but it is beyond our scope. If the purpose is to characterise a lake as a whole then a single *analysis* may suffice, but a single *sample* will do only if the lake is homogeneous. This it rarely is and the first step in any detailed study should be an investigation of spatial (and temporal) variability. As a bare minimum three or more samples from different places may allow this assumption to be tested. If the lake is stratified then more samples will be needed. Inadequate, invalid or even non-existent replication of samples is one of the commonest deficiencies of published work.

If the thermocline moves, both depth and time of sampling must be chosen with great care. If concentration changes with depth but not much with lateral position a 'mixed sample' may be compounded to represent the whole lake. Separate samples from different depths, considered to represent a layer, are mixed in proportion to the volume of each layer.

Even fewer directions can be given about the time and frequency of sampling. Frequency should depend on the rapidity of concentration changes which may depend upon the time of day. During a sunny day the water pH may rise—particularly in poorly buffered waters—due to uptake of CO_2 by plants in the water and the phosphorus concentration for example may decrease. If several lakes are to be sampled it may therefore be necessary to choose a time—perhaps early morning—during which changes are relatively slow. As a general guide the major constituents may be adequately described by samples at 3 month intervals, but minor constituents will probably need to be sampled at least every 7 or 14 days from Spring to Autumn.

1.5.2 Sample collection and preservation

The aim is to transfer the water sample from the lake to the place of analysis with minimal and acceptable chemical change. In a few cases there is no problem because the water is never removed. An example is the electrical estimation of O_2 concentration by apparatus suspended in the water. In most cases however the sample must be removed in a container. With some exceptions there is usually less adsorption on, less diffusion through, and less contamination from, borosilicate glass than from plastics or metal. But for many purposes 'unbreakable' plastic suffices.

Changes in chemical concentration during sampling are roughly inversely related to concentration. Na^+ and Ca^{2+} present few problems but Mo and vitamins far more. Changes are mostly attributable to reaction on the surfaces of the container or to metabolic activity or to change of temperature. These changes may be minimised but no one method of sample treatment between collection and analysis is suitable for all analyses. The analyses to be made must therefore be known before the sample is collected.

Samplers and containers should always be thoroughly cleaned before use, and the material of the sampler should be such that it produces minimal changes in the concentration of analytes either by solution or absorption.

(i) Taking the sample

Sampling disturbs stratification: one should try to reduce disturbance as much as possible. Thus the Ruttner sampler is less suitable for temperature measurement than is the much smaller thermistor thermometer.

Three types of sampler are in common use. The first are open cylinders which can be closed round the sample. The second type are gas or vacuum filled flasks. The third are tube and pump systems.

A. **Open cylinder samplers** (See footnote, p. 27)

Ruttner and Kemmerer types

These have an open tube of 1 to 3 l capacity with a hinged lid at each end. The lids may be closed by a messenger (a weight slid down the suspending cable). The Ruttner tube is made of polymethylmethacrylate (= Perspex = Plexiglas), the Kemmerer of copper.

The lids are held open but they do impede water flow through the tube. Once the desired depth is reached the sampler must therefore be alternately raised and lowered by 25 cm a number of times to flush out contaminant water. This is obviously unsatisfactory but these samplers are nevertheless useful for shallow well-mixed waters.

Friedinger sampler
This sampler is similar to the Ruttner and Kemmerer ones but the lids are held open at 90° (parallel to the axis of the sampler, Fig. 1.5a) and so do not seriously impede water flow. All inside parts are metal free. The capacity is $3\frac{1}{2}$ or 5 litre, and a frame for two reversing thermometers may be added.

B. Gas filled or evacuated flask samplers

Dussart flask
This sampler, Fig. 1.5b, is ingeniously simple and may be home made. A glass flask has a rubber bung through which pass two narrow tubes. One reaches just below the bung, the other to the bottom of the flask. Outside the bung the short projecting ends of the tubes are linked by an 'n' shaped tube or rod attached to the tubes by short pieces of soft rubber tube. A weight attached to the flask causes it to sink. The supporting cable passes round the 'n' shaped rod and then to the

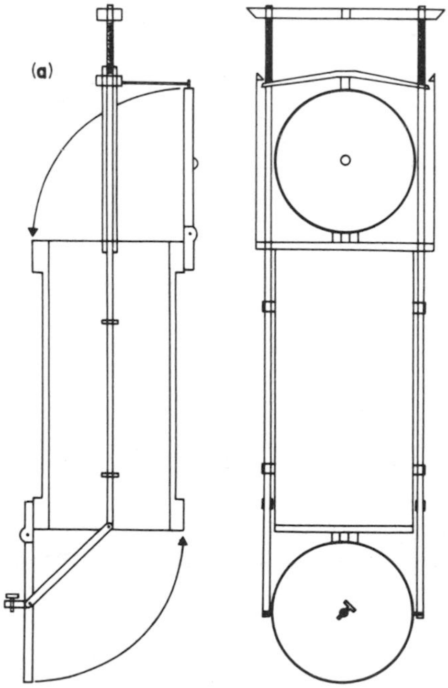

Figure 1.5a. Friedinger 'open cylinder' sampler.

* Obtainable from Hydrobios Apparatebau GmbH, Am Jägersberg 7, 23 Kiel-Holtenau, Germany, and from other suppliers.

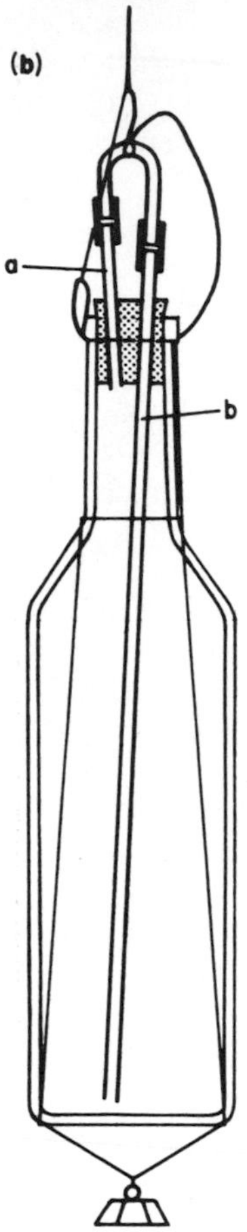

Figure 1.5b. Dussart 'flask' sampler.

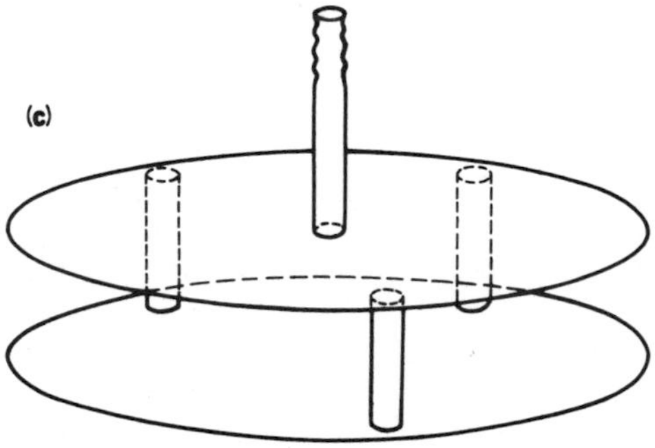

Figure 1.5c. Lower end of 'tube' sampler: the plates ensure that the sample is drawn predominantly from a horizontal layer.

flask neck, but a piece of sewing cotton attached further up the cable and to the flask neck takes the weight. At the desired depth a sharp jerk snaps the cotton and pulls out the 'n' shaped rod. Gas escapes from the flask and is replaced by water.

Valås (flask) sampler
This is similar in principle to the Dussart flask, but the flask closure is magnetic and more complicated. The first jerk opens the flask. The second seals it again. This sampler is therefore specially useful for bacteriological work. It is made by Valås, Otterhall, Göteborg, Sweden.

Watt (1965) flask
Evacuated glass flasks are closed by an expendable glass seal. The messenger smashes the seal.

C. Pump samplers (for example Lund and Talling, 1957)
A weighted rubber or plastic tube is lowered to the desired depth. A pump sucks a continuous stream of water with which the sample flasks are first rinsed then filled. If gases are to be determined the pump outlet is placed at the bottom of the flask and at least three flask volumes pumped through. To restrict the vertical sample dimension the lower end of the tube is attached to the centre of the upper of two horizontal parallel plates of about 10 cm diameter, Fig. 1.5c.

The most suitable type of pump is the peristaltic or finger pump because there is no contact between metal and water. Such pumps act in the same way as squeezing a tube of toothpaste.

The advantages of pumping are that it causes little disturbance and that large samples may be obtained. The 'dead volume' of 10 m of tube with cross section 1 cm^2 is only 1 litre. Pumps are particularly suitable for taking a mixed sample: one simply lowers the tube by regular steps. If the 'layers' are of unequal volume then the volume of sample is adjusted. The collecting flask may be marked in advance for this purpose.

(ii) Filtration
It is convenient to recognise the following fractions in a water sample (though other schemes are possible):

particulate	live
particulate	dead
dissolved	inorganic (generally ionic)
dissolved	organic or bound inorganic

Interconversion can occur during storage. Separation of all four fractions is difficult, but at least the primary separation of particulate from dissolved should be made as soon as practicable. Filtration through a 0·5 μm pore membrane filter under pressure or partial vacuum is the simplest method. It is not suitable if gases, pH or the carbonate system are to be analysed. The 0·5 μm boundary is convenient though unavoidably arbitrary (Olsen 1967). It does seem to correspond with marked change in the properties of the material.

As the sample is filtered the rate of filtration slows because the filter becomes blocked. The area of filter must be chosen to allow sufficient filtrate to collect before the rate becomes unacceptably slow. Glass fibre filters give a higher rate of filtration but the deposit cannot be detached. Membrane filters may release relatively large amounts of organic matter and of compounds of nitrogen and phosphorus. If such release is reproducible then a 'blank' of an equal volume of H_2O may be used to provide a correction. The size of this correction can be reduced by prewashing all the filters with 250 ml H_2O. Glass fibre filters may need pretreating by heating to 500°C or by boiling in a strong oxidising agent (see § 7.1).

The concentration of contaminants may vary widely between batches from one maker, and between different makers.

Sometimes samples must be filtered (and preserved) in the field. A bicycle pump may be used to give positive pressure. Simple, plastic, pistol action pumps may be used to create suction.

Complete separation of living and dead particulate matter is impossible both in theory and practice, but it may be possible to make a useful separation by differential centrifugation and microscopic examination.

Separation of inorganic from organic or bound inorganic solutes can only be made by chemical methods. In most cases the inorganic fraction is estimated, then the organic and bound fractions are converted to the inorganic form and the total estimated. Specific details are given with the methods.

(*iii*) *Preservation*

If the samples cannot be analysed within a few hours of collection then some form of preservative treatment must be given to minimise chemical changes in the sample.

A variety of chemical preservatives has been proposed by i.a. Zobell and Brown, 1944, and Åberg and Rodhe, 1942. Perhaps the four most generally useful are $CHCl_3$, bromine water, and the mineral acids H_2SO_4 or HCl. The acidification is taken to pH 1–2 and has the advantage that it prevents the precipitation of iron and other metals, but it may easily change the state of ions in solution. For trace elements it may be preferable to use HNO_3.

An entirely different method is to freeze the filtered samples in polyethylene bottles. Before analysis the samples must be completely thawed and thoroughly mixed. It occasionally happens that a troublesome precipitate forms during this process, perhaps as a result of the development of high local concentrations during freezing. This method of preservation is not to be trusted if silicates are to be analysed (see § 5.9).

If a variety of analyses are to be made it may be necessary to use more than one method of preservation. The scheme in Table 1.1 is a guide. Detailed instructions are given for particular methods.

Table 1.1. Preservation of samples

Bottle	Analyte(s)	Container Volume	Container Material	Preservative
A	carbonate system (pH, CO_2, HCO_3^-, CO_3^{2-}) rH, SO_4^{2-}, Cl^-, silicate, conductivity	500 ml	glass or thick plastic, dark, no contact with air	2·5 ml $CHCl_3$
A or B	dissolved organic P PO_4–P	as for A but if Fe precipitates then as for B		
B	all others except O_2 and trace elements (nitrogen system, major and other minor constituents, organic C, total P)	500–1000 ml	glass	5 ml 4 M $\frac{1}{2}H_2SO_4$
C	trace elements	1000 ml	glass	5 ml concentrated HNO_3
D	O_2	100–125 ml	glass	See § 8.1.

1.6 METHODS OF SAMPLE CONCENTRATION

It may be that the concentration of an analyte is so low that the accuracy or precision of estimation is unacceptable. A variety of approaches may then be considered. If several analytes are involved then a general concentration procedure may be the best solution. If only one or two are involved then the procedures may be more specific. Probably best of all is to use a more sensitive method, as all concentration procedures are tedious and prone to varying accuracy.

1.6.1 General concentration procedures

All but freeze drying result in high solution concentrations in which unwanted reactions may occur.

(*i*) *Evaporation, using gentle heating*
Boiling may cause loss of sample or change of chemical state or both. The sample may become contaminated by dust but simple apparatus will suffice.

(*ii*) *Evaporation under vacuum using a rotary film or climbing film evaporator*
Special apparatus is needed and the method is impracticable if many samples are to be concentrated. The climbing film evaporator may be made semi-continuous so that it is practicable to concentrate samples of 100 litre or more. In the rotary film evaporator the sample may dry onto the glass.

(*iii*) *Freeze drying*
Expensive apparatus is usually needed and the method is impracticable if there are many or large samples. On a small scale the method is simple. A beaker containing up to 100 ml sample is put in a vacuum desiccator containing concentrated H_2SO_4 as desiccant and a magnetic stirrer follower in the H_2SO_4. The desiccator is put on a magnetic stirrer. Vacuum of less than 1 mm Hg is applied, but once vacuum is established the pump may be turned off. The evaporation of water is so rapid that the whole sample freezes, and the rest of the water sublimes. The sample does not stick to the beaker, and the whole process occurs at low temperature and without high solution concentrations developing. Because convection cannot occur in a vacuum bright light may cause excessively high local temperatures. The desiccator should be kept in dim light.

(*iv*) *Freeze concentrating* (*Shapiro 1961*)
This method is most suitable for samples of more than 1 litre. The sample in a plastic container and with a stirrer paddle operating near the surface is put into a refrigerator. The stirrer paddle should be connected to the motor with a slipping clutch—a piece of rubber tube will do. The sample should freeze from the outside inwards. In some favourable cases pure water separates, and the unfrozen solution becomes concentrated around the stirrer paddle in an egg shaped cavity. This is emptied and rapidly washed with H_2O. Fifty fold concentration may be achieved, but the method is not practicable for large numbers of samples, and is far from quantitative for some elements and concentrations.

1.6.2 More specific methods

(*i*) *Ion exchange*
A large volume of sample may be run through a column of strong acid cation exchange resin in the H^+ form (or Na^+ form if additional Na^+ is tolerable). The

cations from the sample are retained nearly completely by the resin and can be eluted later with about 2 bed volumes of 1M HCl (or NaCl). Using reversed flow of elutant the eluate volume may be reduced to perhaps 0·5 bed volumes. Similar procedures may be used with anion exchange resins in the OH^- or Cl^- form. The method is relatively cheap and can be used for large numbers of samples but there may be problems of varying recovery and interference. The method has the advantage that it may be (and has been) used in the field.

(ii) Solvent extraction

This is commonly used for concentrating the analytic compound; for example, the blue phosphomolybdate may be extracted into hexanol or isobutanol. It may be possible to complex the analyte and extract it before the analysis proper. Many procedures for trace elements use this method.

1.7 CHOICE OF ANALYTICAL METHOD

In general, as urged in § 1.2.4, the analyst should strive to match the method to the problem. This means that the purpose of the work must first be defined then a sampling programme devised. Some preliminary work may be needed to establish the concentrations and variability to be expected.

In general one should try to make accuracy and standard error (as a measure of precision) about the same: imprecise but accurate results are no better than precise but inaccurate ones.

If the problem involves a simple comparison of two fairly stable water bodies of widely different concentration then a relatively crude analytical method may suffice. The answer that one body has 8·4 ± 1·3 concentration units and the other 0·3 ± 0·4 is sufficient for the purpose even though the standard error of the lower concentration overlaps zero.

More commonly, however, decisions are much more difficult to make. In balance studies, for example, it often occurs that one source (or sink) is of high concentration but of small volume and perhaps of short duration whilst another is of low concentration but of high volume.

Suppose in a particular case there are two sources whose concentration and volume during three successive time intervals are:

		Period			Total* amount
		1	2	3	[g]
Source A	Concentration [mg l^{-1}]	1	1	98	
	Volume [m^3]	1	1	1	
	*Amount [g]	1	1	98	100
Source B	Concentration [mg l^{-1}]	5	1	1	
	Volume [m^3]	200	400	400	
	*Amount [g]	1000	400	400	1800
					1900

* Amount calculated from measured concentration and volume.

A positive bias (inaccuracy) of 0·5 mg l^{-1} in the estimation of the concentration of source A on all three occasions would change the calculated total *amount* from

A from 100 to 101·5 g. This is a trivial error in the total of 1900 g. But the same bias in estimation of source B would change the calculated amount from 1800 to 2300 g: a serious error. It might be sufficient to use a crude method to measure the concentration of source A with accuracy $\pm$ 20 mg l^{-1}, but to achieve the same accuracy for the total amount from source B a concentration estimation $\pm$ 0·05 mg l^{-1} is necessary. The accuracy must be inversely related to the *volume* of the source.

Requirements for precision are similarly affected. Using the methods of § 1.2.1 for combining random errors one can show that the random errors are uniformly distributed amongst the components if for every component all the products $Vs_{\bar{c}}$ and $Cs_{\bar{V}}$ are equal, where V = volume of the component, C = concentration, and $s_{\bar{V}}$ and $s_{\bar{c}}$ are the standard errors.

If the volume is large then the standard error of the concentration must be small in proportion. The same applies to the estimate of concentration and the standard error of the volume.

CHAPTER 2

LIGHT AND TEMPERATURE

2.0 INTRODUCTION

Light and temperature are commonly of interest to those who need chemical analyses, and are often measured at the same time that water is sampled. They differ from other analyses because they must, in practice, be measured *in situ*, and because the 'method' is usually simple and obvious but there may be difficulties in deciding what to measure and what equipment to use. At present there are rapid technical advances in this field.

2.0.1 Response time of a measuring device

If a thermometer is transferred from a beaker of water at one temperature to another at a different temperature the rate of change of reading is fast at first but becomes steadily slower. Very often such changes approximate a first order (exponential) decay curve. This is characterised by the 'response time', τ, after which the change is the proportion 1/e, about 37 %, from completion ('e' is the base of natural logarithms). The distance from completion after 1 to 5 response times is:

Response times	1	2	3	4	5
Distance from completion as % of total change	37	13·5	4·9	1·8	0·7

Thus if the response time is 10 s then nearly 50 s are needed for the change to be within 1 % of completion.

The response time for a measuring device does not usually change and need only be determined once. To do so, record the readings at known times after a change and plot the logarithm of the proportion of the change uncompleted against time. A straight line indicates first order response and τ may be read from the graph.

It is useful to know τ because the rate of change of reading may be so slow as to tempt the observer into believing the new steady reading has been reached. It is also valuable to know how rapidly the measuring device is responding to environmental fluctuations, for example on days of alternate sun and cloud.

Continuous sampling systems may have to have τ of only 1 s or so, but rapid response is not always so desirable: fluctuations from second to second are of no limnological significance but make measurement uncertain. The response time can often be increased (equivalent to 'integration' or 'damping' or even, crudely, to 'averaging') by purely electrical means, for example by putting a large capacitor across the instrument output. Detector response time may perhaps be increased too, for example a thermometer may be embedded in a plastic block.

2.0.2 Practical details

Sensors for temperature and light must be suspended over the boat side. For electrical methods it is common to use a single cable for both support and transmission. Take care to avoid overstressing or kinking the cable: a small hand winch mounted on the boat side is valuable particularly for deep measurements. Screened cable (with metal braid concentric with one or more other conductors connected to instrument 'ground' or 'earth') minimises electrical interference. Toroidal rubber 'O' rings allow watertight seals to be easily made and unmade.

2.1 LIGHT

2.1.1 Introduction

The limnologist's most difficult problem when measuring light is to decide what to measure. The answer depends on the purpose for which measurements are needed. We outline here some of the matters which should be considered before any measurement is made. Excellent discussions of many aspects of light and its measurement are given by Westlake (1965), Vollenweider (1971), Monteith (1973), Arnold (1975a), Westlake and Dawson (1975), and Jerlov (1976).

(i) Light, unlike temperature or Ca^{2+} concentration, is not a single homogeneous variable: the light has a spectral distribution. Measuring a single value therefore implies integration over a particular range of the spectrum. Most sensors do this integration automatically but they may give different weight to the various parts of the spectrum. If the spectral distribution of light changes, as it does with depth in a lake or with time of day, then the output from different sensors changes in different ways: the reading of one type may be reduced by much more than that of another type (see Table 2.3).

Apart from this initial spectral integration the limnologist often wishes to integrate over time.

The level of integration must be carefully chosen. Collecting complete spectra at several depths at frequent intervals is very expensive. Integrating separate measurements is also costly. Try to get as much as possible of the integration done automatically by the sensor. The time integral of electrical signals may be obtained electronically or by allowing the current to remove and deposit metal on the plates of an electrolytic cell (voltameter). Such a device is described by Westlake and Dawson (1975).

(ii) Wavelengths in the electromagnetic spectrum span at least 18 orders of magnitude. Light, which is that part of the spectrum visible to the human eye, falls within the approximate limits 400 to 700 nm. The human eye is most sensitive at 555 nm. This range is also roughly that of 'photosynthetically active (or available) radiation' or PAR though the SCOR working group 15 (1965) chose the range 350 to 700 nm for PAR, while others prefer 390 to 700, 710, or 720 nm. Halldal (1967) has shown that light of wavelength down to 310 nm may be effective in photosynthesis in many algae, but light of such short wavelength is relatively rapidly attenuated in lakes (Fig. 2.3). The range from 400 to 3000 nm (0·4 to 3 μm) is commonly called 'short wave' radiation; that from 3 to 100 μm is 'long wave' radiation. Most of the sun's radiation is in the short wave region but (black body) radiation by objects at normal Earth's surface temperature is in the long wave region.

The spectra of some light sources are shown in Fig. 2.1.

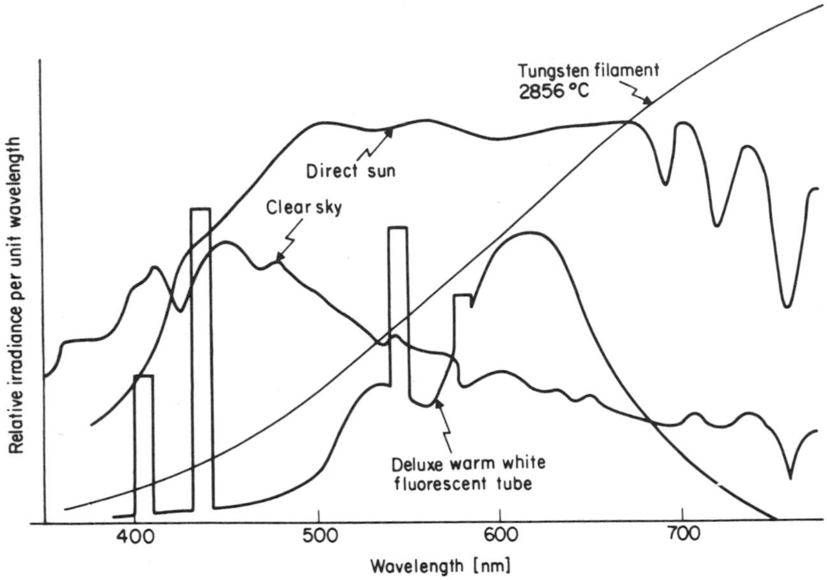

Figure 2.1. Spectral distribution of energy in light from various sources. The units are arbitrary and not comparable between sources. Some absolute comparisons are given in Table 2.2.

(iii) There are three types of 'target' for measurements: *incident* quantities (such as irradiance and photon flux density), *absorbed* quantities (such as the absorption spectrum of chlorophyll), and *effective* quantities (such as photosynthesis action spectra and the widely misunderstood quantity illuminance). The limnologist must choose one or more of these targets and attempt to make the spectral response of the light sensor coincide with that of the target.

Irradiance (irradiant flux density) is a measure of the rate at which energy falls on a surface. Convenient units are W m^{-2} or mW cm^{-2}. Irradiance integrated over time is often reported in units of MJ m^{-2} d^{-1}.

Photochemical reactions (such as photosynthesis) depend in nearly all cases on the number of quanta absorbed. The quantum of light is a photon. It may therefore be useful in some cases to measure not irradiance but photon flux density (PFD). There is a close physical relation between irradiance and PFD because the energy of a photon is inversely proportional to its wavelength. One chemical mole of molecules, atoms, ions or photons contains about $6{\cdot}02 \times 10^{23}$ particles, so the units of PFD are conveniently μmol m^{-2} s^{-1}, sometimes called μEinstein m^{-2} s^{-1} = μE m^{-2} s^{-1}. Some authors prefer quantum m^{-2} s^{-1} with an appropriate multiplier.

Figure 2.2 shows the spectral curves for irradiance and PFD. Table 2.1 gives some equivalences and Table 2.2 some typical values for irradiance.

The incident quantities irradiance and photon flux density are well defined and unique. Absorbed quantities are less well defined and are numerous. The absorption spectrum of chlorophyll *a*, of a whole leaf of *Potamogeton*, or of a suspension of the alga *Anabaena cylindrica* (Fig. 2.2) are examples.

Table 2.1. Some light quantities and units.

Quantity	Irradiance (= Irradiant flux density)		Photon flux density		*Illuminance
Units	W m^{-2} = 0·1 mW cm^{-2}		μmol m^{-2} s^{-1} = μEinstein m^{-2} s^{-1} = 6·02 × 10^{17} quantum m^{-2} s^{-1}		lx = lm m^{-2}
Obsolete Units	cal cm^{-2} min^{-1} $692 = \frac{\text{cal cm}^{-2}\text{ min}^{-1}}{\text{W m}^{-2}}$				foot candle = lumen ft^{-2} $10{\cdot}76 = \frac{\text{ft candle}}{\text{lux}}$
At 555 nm	1·0 W m^{-2}	=	4·15 μmol m^{-2} s^{-1}	=	680 lx

* Illuminance units are given because they have been widely used but illuminance cannot be recommended as a measure in limnology.

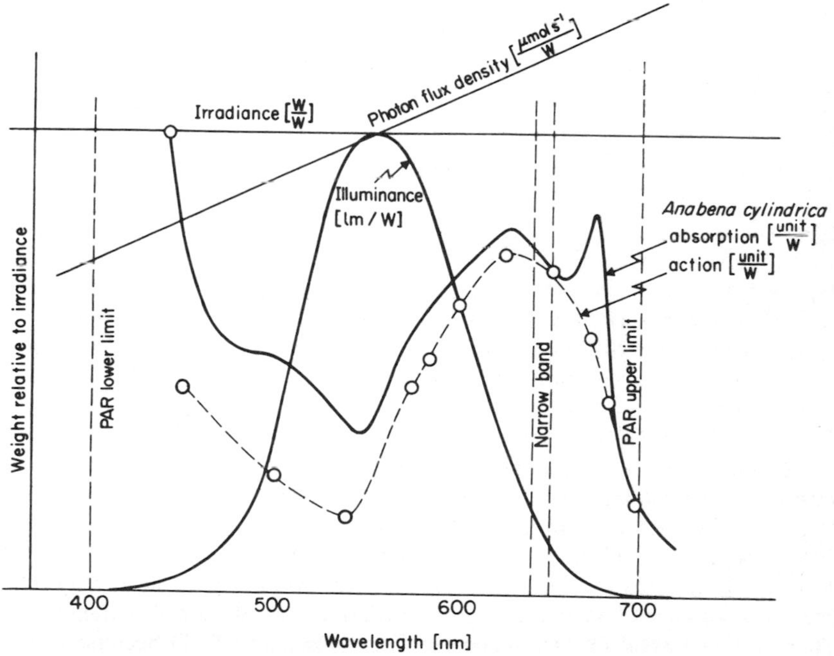

Figure 2.2. 'Targets' for measurement: (a) incident (b) absorbed (c) effective.

(a) Relative weight given to energy flux density as a function of wavelength for irradiance and photon flux density.

(b) Absorption spectrum of whole *Anabaena cylindrica* (Fay, 1970).

(c i) O_2 evolution (photosynthesis) action spectrum of *Anabaena cylindrica* for equal irradiance (Fay, 1970).

(c ii) Illuminance: response of the light adapted young adult human eye.
Both (b) and (c) are illustrative examples of whole classes of possible targets.

Targets may be given specific spectral limits. For illustration a narrow band in the red and the broader 'photosynthetically active (or available) radiation', PAR, are shown. The definition of PAR is not agreed (see text).

Table 2.2. Typical irradiance from various light sources.

Source	Wavelength range nm	Irradiance $W\ m^{-2}$
Outdoor equinoctial midday, no cloud	400 to 3000	600
The same, PAR,	400 to 700	300
Outdoor 52°N summer midday, no cloud, PAR	400 to 700	450
The same, thick cloud		40
50 cm below close-packed warm white fluorescent tubes, PAR	400 to 700	60
50 cm from 1500 W quartz halogen lamp	400 to 15,000	1200
With water screen		350

Equally numerous and usually even less well defined are the effective quantities, for example the photosynthesis action spectrum of *A. cylindrica* (Fig. 2.2). One effective target which *is* well defined is the response of the average photopic (white light adapted) young adult human eye. This is internationally agreed (Commission de l'Éclairage, 1924) and has recognised units. The target quantity is called illuminance, the units are lumen m^{-2} = lx, (Fig. 2.2, Table 2.1). (The symbol for the unit 'lux' is 'lx'). 'Intensity', which is a property of the light source, and 'illumination' are not amongst these quantities or targets and are not usually of interest to the limnologist. They are not commonly measured either, though the terms are commonly mistakenly used for irradiance or illuminance or some measure similar to illuminance.

(iv) The choice of 'target' for measurement depends on the purpose of the work. If the problem involves an energy balance, for example in a study of temperature changes or of evaporation, then irradiance will be needed (and so will other measures not considered in this manual). Irradiance in the PAR range may be best for production studies because transfers to the second and subsequent trophic levels (herbivores, carnivores) are usually expressed in units of energy.

For more detailed work on the photosynthesis of microscopic and macroscopic plants then an absorption target, for example that of chlorophyll *a* or of the whole plant, may be attractive. But this simple notion is complicated by the existence of interactions with other chlorophylls and accessory pigments, by other photosynthetically inactive pigments, by enhancement, and perhaps by photosensitive 'light respiration'. For these reasons an action spectrum target may be considered. Photosynthesis action spectra in light limiting conditions are usually more level than absorption spectra but are known for relatively few plants. The effects of using light of narrow spectral range are often unknown and the experimental irradiance is sometimes very low. The action spectra may well depend on other undefined physiological and environmental variables too. It is clear however that photosynthesis action spectra differ widely. Near the surface of a water body during summer then light is often non-limiting for photosynthesis, so the target would need to be non-linear, and to different degrees for different plants. For general limnological work when there are likely to be varying proportions of algae of different taxonomic groups and perhaps of a variety of vascular plants and varying environmental conditions, then the effective 'target' cannot in practice be defined.

For those whose interest is primarily in photosynthesis there remains the well defined target quantity photon flux density in the PAR range, though it should be remembered that this is an incident and not an effective quantity.

Illuminance as a target can only be justified where the problem involves the human ability to see under water. Even then it is probably incorrect because the sensitivity of the scotopic (dark adapted) eye differs from the photopic one used in the definition of illuminance. The sensitivity of the eye of at least some fish is considerably different from that of the human eye (Protasov 1964, quoted in Arnold, 1975a).

Though illuminance is indefensible as a limnological quantity many published works contain it, or (even worse) illuminance units measured with a sensor which does not have the illuminance spectral weighting shown in Fig. 2.2. To complete the confusion this is often called 'light intensity'. The reason is simple: the selenium barrier layer photocell, with which most of these measurements have been made is cheap, robust and gives a relatively large current which can drive a meter directly. The response curve of this detector is shown in Fig. 2.4. It can be calibrated in illuminance or even irradiance units *for a source with a particular spectral composition*, but the readings will be incorrect if the spectral composition of light falling on the detector changes, as it does if the detector is lowered deeper into water or amongst aquatic vegetation. This may lead to very large errors: Tyler (1973) estimated that 600 or 700% was not unlikely. Table 2.3 gives some guidance.

Table 2.3. Typical interconversions for various sources, quantities and sensors. *These values are illustrative only*: they have been calculated from the curves in Fig. 2.1, 2.2 and 2.4. PAR is taken for convenience between 400 and 700 nm because data are available for that range.

Quantities:	R	= irradiance [Wm^{-2}],
	P	= photon flux density [μmol m^{-2} s^{-1}]
	Aa	= *Anabaena* absorption [arbitrary units]
	Ac	= *Anabaena* action [arbitrary units],
	L	= illuminance [lx]
Spectrum limits	F	= 400 to 3000 nm
	PAR	= photosynthetically active range 400 to 700 nm
Sensors	S	= silicon cell or photodiode, no filter. [Arbitrary units]
	B	= Selenium barrier layer, no filter. [Arbitrary units]

		R/L	P/L	P/R	Aa/R	Ac/R	S	B
Tungsten filament lamp at colour temperature 2856°C	F	0·0538	0·645	12·00	0·043	0·035	1·00	1·00
	PAR	0·0040	0·020	5·01	0·580	0·476	0·82	0·17
De luxe warm white fluorescent tube	F	0·0034	0·017	4·86	0·518	0·453	0·91	0·19
	PAR	0·0033	0·016	4·81	0·541	0·473	0·91	0·15
Loch Croispol 0 m	F	0·0043	0·020	4·69	0·480	0·373	125	26.5
	PAR	0·0040	0·018	4·56	0·521	0·404	125	23·1
Loch Croispol 1 m depth	F	0·0037	0·017	4·61	0·494	0·365	80	14·5
	PAR	0·0035	0·016	4·55	0·518	0·382	80	13·4

The response of the detector, of whatever type it be, must usually therefore be modified by colour filters in such a way that the overall response approximates that of the 'target'.

(v) Light is attenuated in three main ways in water:

(a) reflection at the surface both back into the air and from below back into the water body. The reflectance for direct solar radiation is greater at low than at high sun angle.

(b) scattering by particles, organic and inorganic. This is more effective for shorter wavelengths (blue light) than for longer ones, and particularly effective for particles of dimensions roughly those of the wavelength of light.

(c) absorption by water itself—greater for red than for blue light—and by dissolved coloured matter and suspended organisms. A small fraction (0·1 – 6%) of the light energy absorbed by plants is converted to chemical energy. The rest appears as heat. Plants in the surface layers absorb light of some wavelengths more than of others, so that the light which does penetrate is deficient in just those wavelengths which might be most effective in photosynthesis. A measurement of irradiance or photon flux density will therefore become progressively more misleading as a measure of photosynthetic response. The combined effect of these processes is shown in Fig. 2.3.

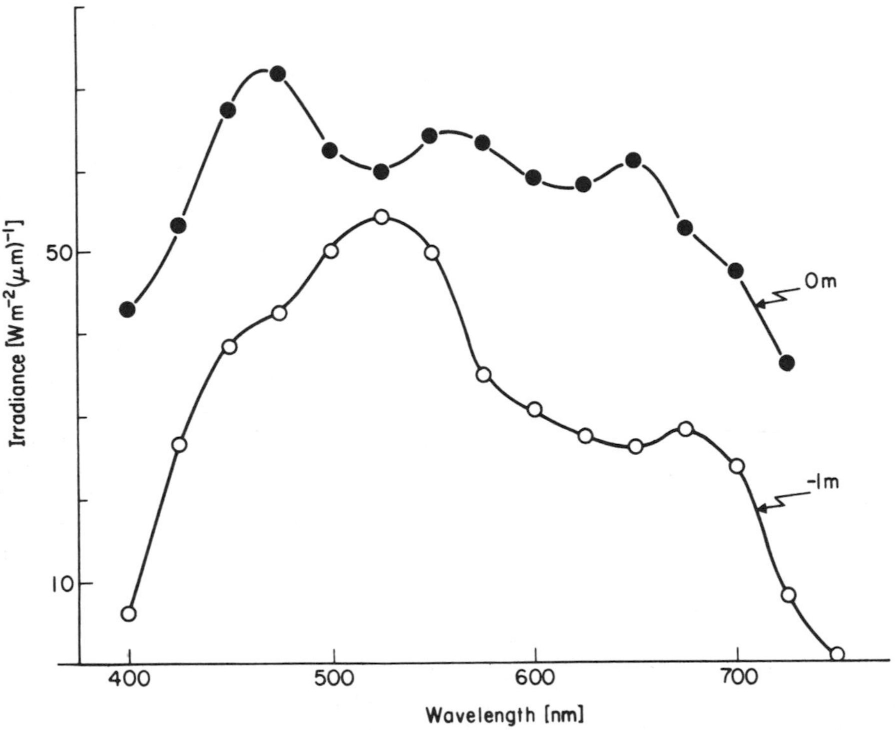

Figure 2.3. Irradiance just below the surface and at 1 m depth in Loch Croispol (calculated with correction of the published units from Spence, 1975). In this example blue and red light are attenuated more rapidly than is yellow.

The characterisation of water bodies by optical properties such as attenuation coefficients, and the directional distribution of light are beyond our scope. Jerlov (1976) discusses these and other matters.

(vi) Detailed attention to all these points would put a prompt halt to many field measurements. This is not our intention. We hope the user of this manual will now be in a position to make informed decisions about what to measure, and to interpret existing measurements made with non-ideal instruments.

2.1.2 General

There follow some notes common to several or all methods.

(i) For some purposes—calculation of heating and evaporation for example—it is necessary to have absolute measurements. For other purposes—some productivity studies for example—then relative measurements may suffice. In yet other cases most measurements may be relative but one of them may be tied to an absolute measurement and this will calibrate the whole set. This is particularly important if a spatial (depth or horizontal) series of measurements is needed because it is unlikely that the light climate at the lake surface will remain sensibly constant for more than a few measurements. Spence (1975) gives an example of the difficulties which ensue. The problem can be largely overcome by using two matched sensors, one of which is kept in a standard position and the other moved to measurement stations. The meter may be switched rapidly between sensors and the moveable sensor recorded as a ratio of the fixed one. (This can even be done electronically with a single integrated circuit if the sensor output is linear).

(ii) Photocells in general are directional. If valid measurements are to be made it is necessary to design the sensor to give the correct 'cosine response': suppose a point light source overhead is moved down along the arc of a circle (as if it were the sun) then the sensor output should be proportional to the cosine of the angle between the light and the vertical. This may be effected with a sheet of translucent Perspex covering the sensor and acting as a light diffuser. The design of this diffuser is critical (Smith, 1969). Residual deviations of 3–5% may be expected with a good design, but a bad design may cause undefined errors of 1000% (10 fold) or more.

(iii) Measurements near the surface are particularly liable to error as a result of unavoidable multiple reflection between the sensor and the water surface and should not be attempted at depths less than 10 cm. They are also particularly affected by the shadow of the observer's boat. Sensors should be suspended as far from the boat as practicable and should be mounted so that they remain horizontal.

(iv) Many sensors are linear at low irradiance but become non-linear at the levels commonly reached near the surface at midday. In such cases use a neutral filter (grey tinted glass, black nylon gauze, or stainless steel wire gauze, which attenuates light at all wavelengths by equal proportions) to reduce the irradiance to the linear range. The transmittance of the neutral filter must be known.

(v) Reflectance from the surface of a sensor in water is different from that in air (Westlake, 1965). To correct for this two sensors are first calibrated in air. One is kept in air and the other placed under wave-free water at successively greater depths at 10 cm intervals. The results are plotted as log response against depth and the line extrapolated to zero depth to give the apparent subsurface response. The expected (dry) response (100%) minus the expected surface reflectance (say 10%) gives the expected sub-surface response (then 90%). [The expected surface reflectance is got using sun angle, and cloud cover with Anderson's (1952) tables, or, approximately, from Table 2.4]. If the apparent subsurface response is 70%, then

Table 2.4. Reflectance from still water surface [%].

Solar elevation (angle to horizontal)	50°	40°	30°	20°	15°	10°	5°
Direct sunlight	2	3·5	6	13	21	35	58
Diffuse sky light	7	9	11	13	14	15	17

the immersion correction factor is 90/70. Smith (1969) and Westlake and Dawson (1975) describe this procedure.

(vi) Absolute calibration over a range of wavelengths is far from easy. Arnold (1975a, 1975b) describes one procedure. For many purposes it is sufficient to calibrate irradiance against a stable thermopile: for example the Kipp or the Eppley pattern.

LEVEL I

2.1.3 Secchi disc depth

Principle

The depth at which a target (Secchi) disc is just visible is recorded. This depth is partly dependent on the light flux but mainly on the optical properties of the water. The method is included under 'light' for convenience.

The method is quick and cheap if the limnologist is at the lake for another purpose but is very dependent on the observer, the time of day, the method of observation and on weather conditions.

In favourable conditions (calm water, bright light) irradiance at the Secchi disc depth is about 15% of that just below the surface. The euphotic limit—the depth at which irradiance is 1% of that at the surface—is about $2\frac{1}{2}$ times the Secchi disc depth, and the vertical attenuation coefficient ε_{v} is roughly 1·9/Secchi disc depth.

These calculations assume however that the water column is homogeneous, that the Secchi disc depth is related to irradiance and in particular to 15% of the surface value, and that vertical attenuation is the dominant process. The truth of these assumptions can rarely be tested: if equipment is available to do so then the Secchi disc is superfluous. The calculations should therefore be treated as rule of thumb estimates subject to error by at least a factor of 2. The method is much better than nothing however. Correlation with other measures is discussed by Vollenweider (1971).

Secchi disc depths of 0·1 to 40 m are credible.

Apparatus

A disc 25 to 50 cm diameter, painted matt white. Discs with black quarterings or octants are sometimes used, though some workers doubt the advantages of such patterned discs.

Method

Make measurements during the middle of a sunny day. Take the average of the depth at which the disc just disappears and just reappears. The results may be more reproducible, though different, if diving glasses or an observation box dipping into the water are used. This is particularly the case if there are a lot of small waves. Polarizing sunglasses may increase the Secchi disc depth remarkably!

There is little agreement about the size and pattern of markings on Secchi discs, so results are even more difficult to compare than need be.

Coloured Secchi discs (blue, green, red) may be helpful, though they have been little used. The spectral composition of light and the spectral sensitivity of the human eye complicate interpretation.

2.1.4 Total incident radiation

Principle
Photochemical reactions, for example the polymerisation of anthracene (Dore, 1958) or the decomposition of oxalic acid photocatalysed by uranyl sulphate (Bucholz, 1805, Niepce de Saint-Victor et Corvisart, 1859, Atkins and Poole, 1929, Westlake, 1966) have been used to integrate irradiance over time. Such methods are excessively dependent on a limited part of the spectrum and have correspondingly little value. We give no further details here.

LEVEL II

2.1.5 Short wave irradiance by thermopile (solarimeter)

Principle
Radiation falls on a black and white surface. The black area becomes hotter than the white and the small temperature difference is measured by a series of thermocouple junctions which alternate between the black and white areas.

The combined potential differences are a linear function of irradiance (in a well designed instrument). The output, which at noon outside in Spring is about 5 – 15 mV depending on the instrument, may be recorded by potentiometer or it may be used to drive a sensitive ammeter. The resistance of the ammeter should match that of the sensor—about 10 to 100 Ω. The output may be integrated electrically. If only the total incident energy density is needed this is strongly recommended. Integration of charts is slow and costly.

A well designed sensor has a nearly level response between 400 and 2700 nm if glass covered and up to 15 μm at least (with a few opaque bands) if polyethylene covered. Unfortunately it is not very sensitive and is of greatest use as a surface standard.

Commercial instruments of the Kipp and of the Eppley pattern are available. A less accurate integrating unit is available from Lintronic.

These instruments are expensive, but ones of lower accuracy may be made cheaply. Designs have been published by Monteith (1959) and a development of his design by Szeicz (1966). All these are intended for aerial use, and are of value to limnologists simply as local standards.

In air and normal daylight one may usually assume that light or PAR is 0·50 times the 400 to 3000 nm irradiance (Szeicz, 1975), but it may be preferable to use two instruments one of which has a filter (for example RG 8, Fig. 2.6) which cuts out all wavelengths below 700 nm. PAR is then got by difference. The glass dome provides the cut off at 400 nm. A more expensive solution is the Calflex BI/KI combined interference filter (Fig. 2.6) which cuts out all wavelengths greater than 700 nm and therefore needs only a single solarimeter.

Method
Follow the maker or designer's instructions. The outer dome must be kept clean. For critical work the sensor should be recalibrated every two years (at least). Home made sensors probably need repainting every year and recalibrating every month or two.

It is useful to know the response time of the sensor (§ 2.0.1). For example Szeicz's solarimeter has a response time of 30 sec. Response to sudden change in irradiance will not be substantially complete in less than 2 minutes, and short term fluctuations lasting less than 20 seconds will be markedly damped.

2.1.6 Other light sensors

The sensitivity of detectors other than thermopiles is markedly wavelength dependent (Fig. 2.4). No sensor has the spectral response needed for measurements of

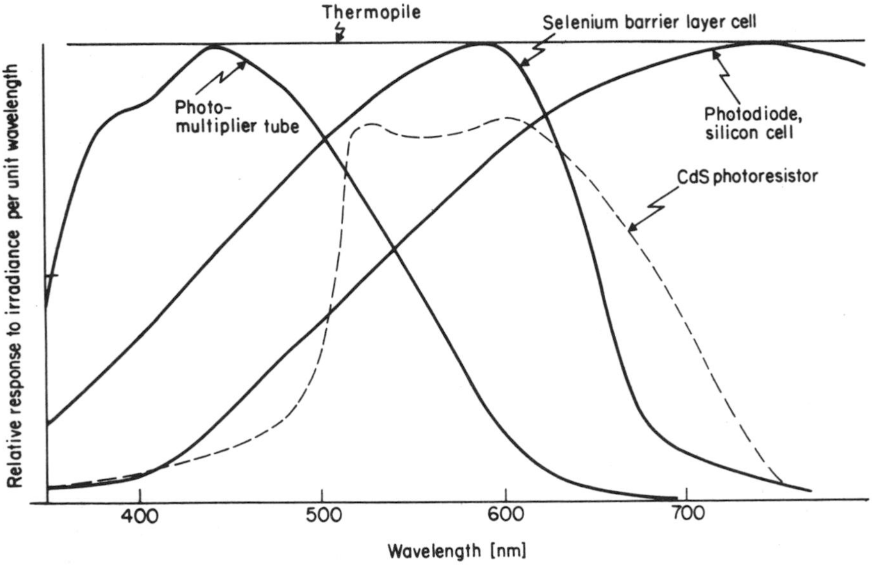

Figure 2.4. 'Detectors': spectral sensitivity of various light detectors. These are typical for the type but *there is variation within types* i.e. do not assume that the curve here refers to your detector of this type. This is particularly true for photomultiplier tubes.
The units are not comparable between detectors.

photon flux density. The selenium barrier layer photocell—the most commonly used detector—approximates that needed for illuminance measurements (Fig. 2.2) and can be made to do so even more closely using a filter, but this is of little use to the limnologist (§ 2.1.1). On the other hand the electrical characteristics of these detectors are much more favourable than are those of thermopiles. For all such detectors two solutions are possible:

(a) use a filter or combination of filters which converts the spectral response to that needed for irradiance (flat) or photon flux density measurements (Fig. 2.2). The design procedure is illustrated in Fig. 2.5. Another design giving both irradiance and photon flux density measurements has been published by Jerlov and Nygard (1969) and a third (available commercially from Lambda Instruments) by Uphoff and Hergenrader (1976).

(b) use filters which isolate narrow spectral bands and measure, and calibrate, with each separately. The properties of interference filters are excellent but the filters are rather delicate and their properties may be changed if the filter becomes wet. The spectral characteristics of some filters are shown in Fig. 2.6. In particular Schott make a series of glass filters with cut off separated by about 20 nm from each other and covering the whole visible spectrum. Evans (1969), who first drew attention to these, gives characteristics. By using two together it is possible to delimit a narrow band in the spectrum. Figure 2.6 gives an example. Reconstructing the complete spectrum is not easy however (Edwards and Evans, 1975).

The CdS detector differs from others in that the property which varies with irradiance is electrical conductance. Its spectral response is not easily transformed to irradiance or photon flux density and it has not been much used in limnology. But it is possible to use a conductance bridge for measurements and there may be circumstances in which this would be useful.

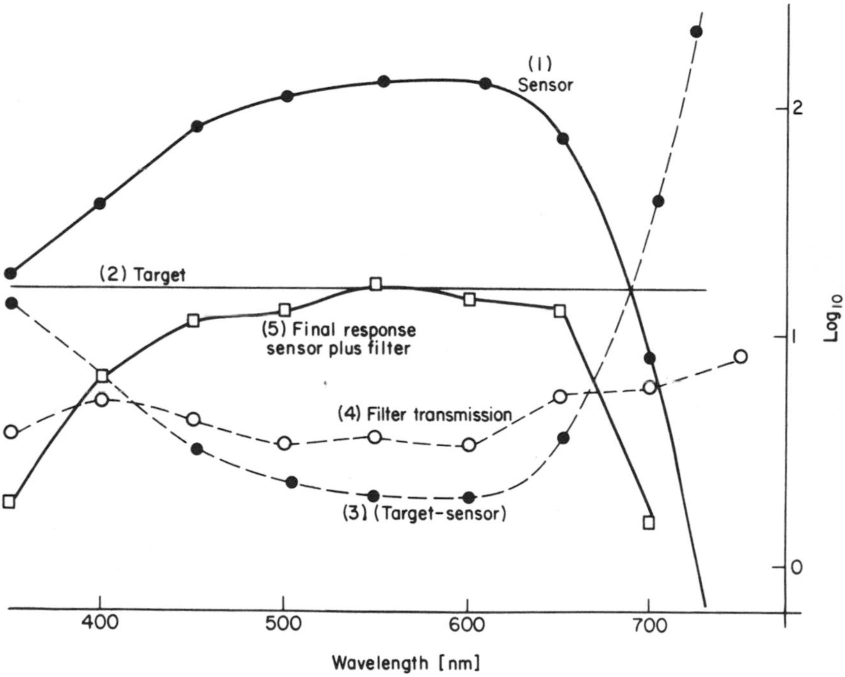

Figure 2.5. Designing a sensor to measure irradiance in the PAR range. The example is that of Powell and Heath (1964) modified by Westlake and Dawson (1975), though the logarithmic plot used here is not theirs. First the sensor—a selenium barrier layer photocell (Megatron B)—is chosen and the spectral response (1) graphed with a logarithmic scale. Next the target—irradiance (2)—is graphed, again on a logarithmic scale. The vertical position is unimportant. Next the vertical distance (3) between target and sensor curves is plotted. This distance, a difference of logarithms, is the ratio target/sensor response and is the curve which must be *paralleled* by the filter transmission, again on a log scale. The filter curve (4) is for the combination of Cinemoid 'Steel Blue 17' and 'Pale Salmon 53' filters (Strand Electric). The overall response of sensor and filters is shown as (5), which has been shifted vertically to aid comparison with the target (2).

The Si photocell is potentially useful. If the output is fed through a low (10–50 Ω) external resistor the response is linear, but fed through a high resistor the response is logarithmic.

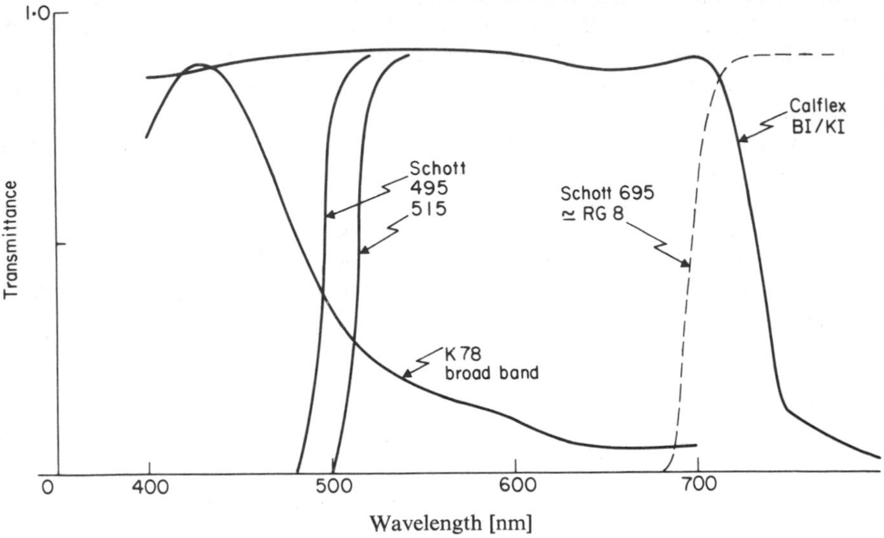

Figure 2.6. 'Modifiers': absorption spectra of various filters.
There are a very large number of types of filter. The makers Schott, Kodak and Chance all produce a wide range of types whilst smaller specialist makers produce some filters with very useful properties.

Photomultiplier tubes, which are now used in oceanography, are the most sensitive of all detectors but need high voltages and retain a memory of saturation, which occurs at very low irradiance. They may be able to detect a black cat in a dark room but are not a practical proposition for the non-specialist.

Much of the older apparatus used by limnologists has an unsuitable spectral response. Even so it may, if used with discrimination, be better than Secchi disc measurements. The conversion factors shown in Table 2.3 on page 39 may be helpful.

LEVEL III

2.1.7 Irradiance spectrum

Principle

Spectroradiometers are able to measure irradiance (or photon flux density or both) within a narrow (5 to 20 nm) band in the visible spectrum. They can scan the spectrum and it is thus possible to construct the complete irradiance spectrum. The instruments should be checked against a certified local standard source (Arnold, 1975b). Relative readings at different wavelengths are more reliable than absolute values.

Remote sensors coupled by fibre optic leads are used in some designs, for example Isco. In others, for example Techtum QSM 2500, the sensor is connected by wires which may be 100 m or more long.

Given the irradiance spectrum then any other quantity—photon flux density or an action spectrum weighted quantity similar to illuminance—can be calculated. The instruments are expensive.

2.2 TEMPERATURE

Excellent reviews of temperature measurement are given by Mortimer (1953) and by Mortimer and Moore (1953, revised 1970).

Apparatus

There are four types of apparatus in common use. Their characteristics are summarised in Table 2.5. In brief they are:

(i) Reversing thermometer: after the thermometer has reached equilibrium at the required depth a messenger weight is sent down the supporting cable and causes the thermometer to invert in its frame. This separates the Hg thread from that in the main bulb, thus 'preserving' the reading while the thermometer is hauled up. The apparatus can be accurate, but is expensive and can be used for only a single measurement at each immersion. Reversing thermometers are sometimes attached as standard equipment to open cylinder samplers (§ 1.5.2).

(ii) Thermocouples are very easy to construct but need a reference junction at known temperature—for example a vacuum flask of lake water. They generate a very small potential which is not easily amplified with accuracy and is liable to disturbance by electrical pick-up (Long, 1968), but are excellent for measuring small temperature differences.

(iii) Thermistors have a large negative temperature coefficient of resistance. They are fairly cheap but need fairly frequent recalibration. Their response is non-linear but may be made nearly linear over a limited range by a fixed resistor of equal magnitude in parallel. This reduces the sensitivity however.

(iv) Nickel (or platinum) resistance elements are moderately expensive, though they are fairly easily made (Long, 1968). They are very accurate and stable. The measurement with thermistor or nickel wire element is of resistance.

Table 2.5. Typical characteristics of four types of thermometer.

	Mercury in glass reversing thermometer	Thermocouple (copper-constantan)	Thermistor	Nickel wire
Precision °C	0·01	0·1	0·05	0·01
Measurement	length	electrical potential	electrical resistance	electrical resistance
Magnitude of output		500 μV	1 kΩ to 100 kΩ	100 Ω
Change per °C	5 cm	40 μV	4% (1%)*	0·6%
Linearity	good	good	poor (modest)*	good
Stability	high	moderate	reputedly low, but may be high	high
Typical response time†	20 s	0·2 s	2 s	30 s
Operation	single measurement	continuous	continuous	continuous
Sensor cost	very high	low	moderate	high
Meter cost	none	moderate to high	moderate‡	moderate

* a resistor in parallel with the thermistor improves linearity but decreases sensitivity.
† see definition in § 2.0.1. Actual value is very dependent on construction.
‡ it may be possible to use a conductance bridge: see text.

Measurement of resistance can in principle be made with a conductance bridge, though the requirements of precision, accuracy and sensitivity may be beyond the capability of the cruder battery operated instruments. A suitable self balancing bridge is described by Moore (1976). For nickel wire thermometers the range of a conductance bridge will usually be wrong too, but thermistors can be chosen with resistance in the correct range, usually about 1 kΩ to 10 kΩ. Using such thermistors there is less likely to be (unrecognised) trouble from dirty contacts. Low resistance elements should always be connected through gold flashed connectors. In resistance thermometers a current flows through the detector and causes self-heating, but in water, with its high thermal capacity this is not likely to be a problem unless the applied voltage is high. One should try to disperse less than 0·1 mW from the sensor.

The sensor should be shielded from direct solar radiation: in water a simple 'hat' is sufficient, but even this is unnecessary on dull days or more than a few metres below the surface.

Choice of thermometer depends on purpose and availability of equipment. For general purposes a frequently calibrated thermistor is probably the most useful. If the occasional more accurate measurement is needed and a reversing thermometer is available it will do very well. For the most accurate work, needing discrimination to 0·01°C, as for example in some studies of stratification, then nickel wire thermometers may be preferred.

Apparatus used for other measurements may include a thermistor as part of a temperature compensation circuit. It may be possible to use the same thermistor for temperature measurement. An example is the Mackereth cell for O_2 estimation. If this is done it is necessary to be sure that the thermistor is measuring water temperature and not just the cell temperature.

CHAPTER 3

CONDUCTIVITY, pH, OXIDATION–REDUCTION, POTENTIAL, ALKALINITY, TOTAL CO_2, ACIDITY

For theoretical considerations see chapter 9.

3.1 CONDUCTIVITY

Introduction

Electrical conductance is the reciprocal of electrical resistance. It is a measure of the ability of a conductor to convey electricity. Hydraulic conductance is a measure of the ability of a conduit to conduct a fluid and thermal conductance of a substance to conduct heat. All three may be used by limnologists: the meaning of conductance (unqualified) will usually be clear from the context. Here we intend electrical conductance.

The conductance of water is measured between two rigidly mounted parallel plates of platinum or carbon and is directly proportional to plate area and inversely to plate separation. When the results are expressed as if for plates of 1 cm^2 separated by 1 cm then the measure becomes independent of the measuring system and is a property particular to the water sample. This measure is the electrical conductivity. The unit is S cm^{-1} (or Ω^{-1} cm^{-1}), where the 'S' commemorates Siemens. The conductivity of most fresh waters is so low that the unit μS cm^{-1} is commonly used. The older name 'specific conductance' and unit μmho cm^{-1} are now obsolete.

In practice, for a given plate configuration, the factor by which conductance must be multiplied to give conductivity is measured empirically using a standard electrolyte solution of tabulated conductivity. This factor is called the 'cell constant'.

Differences in the conductivity of water samples are mainly a result of differences in the concentration of charged solutes and to a lesser extent in the nature of the solutes and of temperature. Conventionally results are reported at 20°C or 25°C. The higher the concentration of charged solutes the higher the conductivity. For water with pH above about 4·5,

$$c \simeq 0{\cdot}01\ \kappa$$

where κ = conductivity expressed in μS cm^{-1}.

c = sum of concentrations of positively and negatively charged ions expressed in mmol l^{-1}.

A more accurate calculation which may be used to check an analysis is given in § 4.10.

Apparatus for the measurement of conductivity is sold by many makers. The instruments all consist of two distinct parts.

(a) A conductance cell containing a pair of rigidly mounted 'electrodes', usually made of Pt with a coating of Pt-black, or of carbon.

(b) An instrument for measuring the electrical conductance (or resistance) between the electrodes of the cell.

Most instruments consist of a low voltage alternating current generator feeding a Wheatstone bridge network, of which the cell forms one arm.

Battery operated apparatus may be useful for field measurements of changes with time, or to detect chemoclines (changes of concentration with depth).

LEVEL I

3.1.1 With field apparatus

Apparatus

For field measurements specially designed portable, battery operated apparatus is necessary. Such apparatus may easily be made by those with an elementary knowledge of electronics. The conductance cell may be bought separately. The cell constant should be measured in the laboratory using a solution of KCl of known concentration (see Level II, § 3.1.2).

Conductance cells containing electrodes of impervious carbon are available and are particularly suitable for monitoring and field studies. (For example from Electronic Switchgear). A thermometer covering the range —5° to 30°C is also necessary.

Procedure

Clean the 'electrodes' and, if necessary, replatinize them (§ 3.1.2). Rinse the cell well with the sample to be measured, finally filling the cell. Measure the conductance (or resistance) of the sample in accordance with the instructions supplied by the manufacturer. Measure the water temperature.

Calculation

See § 3.1.2.

If the conductivity was measured at a temperature other than 25°C, an approximation to the value at 25°C may be obtained by multiplying by a factor obtained from Table 3.1.

Table 3.1. Factors for converting conductivity of water to values at 25°C (based on 0·01 M KCl and 0·01 M $NaNO_3$ solutions).

°C	factor	°C	factor	°C	factor
32	0·89	22	1·06	12	1·30
31	0·90	21	1·08	11	1·33
30	0·92	20	1·10	10	1·36
29	0·93	19	1·12	9	1·39
28	0·95	18	1·14	8	1·42
27	0·97	17	1·16	7	1·46
26	0·98	16	1·19	6	1·50
25	1·00	15	1·21	5	1·54
24	1·02	14	1·24	4	1·58
23	1·04	13	1·27	3	1·62

If the pH of the sample is below 4·5 it may be desirable to subtract the contribution of H^+ to the conductivity because the molar conductivity of H^+ is 7 to 10 times that of most other cations. The same is true for OH^- above pH 10. Table 3.2 gives the necessary constants. The results are then usually called 'corrected conductivity' or 'reduced conductivity'. Such 'corrections' are only approximate.

Table 3.2. Conductivity attributable to H^+ or OH^- at 25°C.
A shift of 1 unit in pH causes a tenfold change in conductivity. The table may therefore be extended easily; for example at pH 3·3 the 'correction' is 180. The values are progressively more unreliable outside the range shown.

pH	3·5	3·6	3·7	3·8	3·9	4·0	4·1	4·2	4·3	4·4	4·5
H^+ $\mu S\ cm^{-1}$	114	91	72	58	45	35	29	23	18	14	11·4
pH	9·5	9·6	9·7	9·8	9·9	10·0	10·1	10·2	10·3	10·4	10·5
OH^- $\mu S\ cm^{-1}$	6·3	8	10	13	16	20	25	32	40	50	63

LEVEL II

3.1.2 With laboratory apparatus

Apparatus

There are several satisfactory commercial models. In addition one needs a large waterbath, with racks in which tubes containing samples and standards may be kept at a known constant temperature. Large borosilicate test tubes are convenient for holding the samples. It is essential that the diameter of the tube be at least twice the diameter of the cell. Soda glass tubes are unsuitable for soft water samples because appreciable changes in conductivity may result from solution or exchange of ions from the glass.

Reagents

H_2O

For conductivity work the H_2O must be freshly distilled then allowed to cool protected from the CO_2 and sulphur oxides in air. The conductivity of this water should be less than about 2 $\mu S\ cm^{-1}$. The H_2O may have to be passed through an ion exchanger to reduce the conductivity to this level.

KCl, 0·0100 M

Dissolve 0·7456 g of KCl (A.R., dry) in freshly boiled (CO_2 free) H_2O with a conductivity not greater than 2 $\mu S\ cm^{-1}$, and make up to 1000 ml. Store in a glass

Table 3.3. Conductivity of KCl solutions at 25°C.

Concentration M	Conductivity $\mu S\ cm^{-1}$
0·0001	14·94
0·0005	73·90
0·001	147·0
0·005	717·8
0·01	1413·0
0·02	2767
0·05	6668
0·1	12900
0·2	24820

stoppered borosilicate bottle. This standard is satisfactory for most samples when using a cell with a constant between 1 and 2. Stronger or weaker standards may be needed; see procedure and Table 3.3. When preparing very dilute standards it is essential to use freshly prepared conductivity water.

Preparation and care of conductance cells

In normal use the cell should be rinsed with H_2O before storage. If salts or organic matter have dried onto the surfaces, the cell should be immersed in chromic acid (§ 1.2.5) for about an hour.

Occasionally the platinum cells must be replatinized. Declining accuracy may be a sign that this must be done. Clean the cell in chromic acid (§ 1.2.5). Platinize by immersing the cell in a solution of 1 g of chloroplatinic acid ($H_2PtCl_6{\cdot}6H_2O$) and 10 to 20 mg of $Pb(CH_3COO)_2$ in 100 ml of H_2O. Alternatively use 0·3 g of $PtCl_4$ and 10 to 20 mg $Pb(CH_3COO)_2$ in 100 ml of 0·025 M HCl. Connect *both* plates to the negative terminal of a 1·5 V cell. Connect the positive side to a Pt-wire dipped into the solution. Continue the electrolysis until both electrodes are coated; only a small quantity of gas should be evolved (corresponding to a current density of about 10 mA cm^{-2}). Keep the solution for later use.

Procedure

The water should, if possible, be maintained at 25°C, although any controlled temperature between 15°C and 30°C may be used, as the conductance of the KCl standard will vary with temperature to nearly the same extent as that of the samples. The conductivity cell should be stored in H_2O at the temperature of the waterbath. If the conductivity is to be used to give a check on a complete analysis (§ 4.10.2) first measure the conductivity approximately. Calculate the dilution necessary to make the conductivity about 100 $\mu S\ cm^{-1}$. For example if the conductivity of the sample were 600 $\mu S\ cm^{-1}$ then a sixfold dilution should be suitable.

Make this dilution accurately using conductivity H_2O.

Place four tubes of standard KCl solution (see Table 3.3) in the waterbath. Place two tubes of each sample to be measured in the waterbath, and allow 30 minutes for thermal equilibrium to be established. A KCl solution should be chosen with a conductivity differing by a factor of less than 5 from that of the samples (see Wilcox, 1950).

Rinse the conductivity cell with the contents of three of the tubes of KCl solution in turn, measure the conductance (or resistance) of the fourth solution and record the results.

Next rinse the cell with one tube of the first water sample, ensuring that the rinsing is thorough, and measure the conductance (or resistance) of the solution in the second tube. Proceed in the same way with the other samples.

Calculation

First calculate the cell constant, K[cm^{-1}] using the measurements with standard KCl:

$K = R_m \kappa_t$ (meter calibrated for resistance)

or $K = \kappa_t / G_m$ (meter calibrated for conductance)

where R_m = measured resistance [$M\Omega = 10^6\Omega$]

G_m = measured conductance [$\mu S = 10^{-6} S = 10^{-6}\ \Omega^{-1}$]

κ_t = conductivity of the standard KCl [$\mu S\ cm^{-1}$] (Table 3.3)

The conductivity of the sample, $\kappa_s[\mu S\ cm^{-1}]$, is given by:

$\kappa_s = K/R_s$ (for resistance)
or $\kappa_s = KG_s$ (for conductance)

where R_s = resistance of sample [MΩ]
G_s = conductance of sample [μS]

For example, suppose the measured resistance for 0·01M KCl is 0·900 kΩ (= 9×10^{-4} MΩ). Then the cell constant is:

$$K = 9 \times 10^{-4} \times 1413 = 1{\cdot}27\ cm^{-1}$$

Given the conductivity $\kappa[\mu S\ cm^{-1}]$ the total concentration of charged solutes [mmol l^{-1}] may be estimated approximately as 0·01κ or even more approximately [mg l^{-1}] as 0·75κ. The reliability of this last calculation is no better than ±50% (see § 9.1).

3.2 pH

pH may be measured by a colorimetric method (using indicators) or by an electrical method. The colorimetric method requires little capital expenditure but is less convenient, less accurate, and needs more time for a determination than the electrical method.

There may be local circumstances which force the limnologist to use the colorimetric method, and for this reason suitable buffer solutions and indicators are described in § 3.2.4, but colorimetric methods § 3.2.1 and § 3.2.2 have been omitted from this edition.

LEVEL I

3.2.3 Electrometric, with a battery operated field pH meter

Apparatus

Battery operated pH meters may be bought. As instructions are given by the manufacturer no procedure is detailed here. See § 9.5 for the effects of temperature.

Reagents and Procedure

See § 3.2.4

The following points should be noted.

The accuracy in normal use is probably ± 0·1 of a pH unit, but can be improved by checking against buffers of a known pH value close to that of the sample.

It is particularly important with portable pH meters to check that the scale factor is correct. This may be done by standardising with one buffer solution and then checking that the correct reading is obtained (without further adjustment of the controls) on a second buffer solution with pH 2–3 units different from the first.

When the apparatus is not in daily use batteries which are not leakproof are best removed. The apparatus must be kept dry.

LEVEL II

3.2.4 Electrometric, with a laboratory pH meter

Apparatus

Very many types of pH meters are commercially available. For most limnological work an ordinary laboratory pH meter with an accuracy of ± 0·02 pH units will be sufficient. Only the investigator can decide when greater accuracy is needed. The cost will, however, increase more rapidly than the accuracy.

Two electrodes are necessary with the apparatus; a glass electrode (see § 9.9) and a calomel electrode. A KCl salt bridge links the calomel electrode to the sample. Combined glass and reference electrodes, including the salt bridge, can be obtained. The reference junction is usually Ag-AgCl.

See § 9.5 for the effects of temperature.

Reagents

A Buffer solutions—*suitable as standards for calibration of pH meters.*

Potassium hydrogenphthalate, 0·05 M
Dissolve 10·2 g of K hydrogen phthalate in H_2O and dilute to 1000 ml.

Phosphate buffer, 0·05 M.
Dissolve 3·40 g of KH_2PO_4 and 4·45 g of $Na_2HPO_4 \cdot 2H_2O$ in H_2O and dilute to 1000 ml.
This solution is 0·025 M in $H_2PO_4^-$ and in HPO_4^{2-}.

Borax buffer, 0·01 M.
Dissolve 3·81 g of $Na_2B_4O_7 \cdot 10H_2O$ in H_2O and dilute to 1000 ml.

pH of buffer solutions A at various temperatures.

Temperature °C	Phthalate	Phosphate	Borate
0	4·01	6·98	9·46
5	4·00	6·95	9·39
10	4·00	6·92	9·33
15	4·00	6·90	9·27
20	4·00	6·88	9·22
25	4·01	6·86	9·18
30	4·01	6·85	9·14

B Buffer solutions—*suitable for maintaining intermediate pH values.*
Acetate Buffer (Walpole, 1914).

CH_3COONa, 0·20 M.
Dissolve 27·2 g of $CH_3COONa \cdot 3H_2O$ in H_2O and dilute to 1000 ml.

CH_3COOH, 0·20 M.
Dilute 12 ml glacial CH_3COOH to 1 litre with H_2O.
Standardise against NaOH, with phenolphthalein as indicator, and dilute to 0·20 M

Range of pH values which may be obtained by mixing CH_3COONa and CH_3COOH solutions

pH at 18°C	0·2 M CH_3COONa (ml)	0·2 M CH_3COOH (ml)
3·6	0·75	9·25
3·8	1·20	8·80
4·0	1·80	8·20
4·2	2·65	7·35
4·4	3·70	6·30
4·6	4·90	5·10
4·8	5·90	4·10
5·0	7·00	3·00
5·2	7·90	2·10
5·4	8·60	1·40
5·6	9·10	0·90
5·8	9·40	0·60

This buffer is often colonised by micro-organisms which metabolise the acetate and change the pH. $HgCl_2$ may prevent their growth.

Phosphate Buffer after Sorensen (Gomori, 1955).

Na_2HPO_4, 0·2 M.
Dissolve either 35·6 g of $Na_2HPO_4 \cdot 2H_2O$ or 71·6 g of $Na_2HPO_4 \cdot 12H_2O$ in H_2O and dilute to 1000 ml.

NaH_2PO_4, 0·2 M.
Dissolve either 27·6 g of $NaH_2PO_4 \cdot H_2O$ or 31·2 g of $NaH_2PO_4 \cdot 2H_2O$ in H_2O and dilute to 1000 ml.

Range of pH values which may be obtained by mixing these phosphate solutions with H_2O .

pH	0·2 M Na_2HPO_4 (ml)	0·2 M NaH_2PO_4 (ml)	H_2O (ml)
5·8	8·0	92·0	100
6·0	12·3	87·7	100
6·2	18·5	81·5	100
6·4	26·5	73·5	100
6·6	37·5	62·5	100
6·8	49·0	51·0	100
7·0	61·0	39·0	100
7·2	72·0	28·0	100
7·4	81·0	19·0	100
7·6	87·0	13·0	100
7·8	91·5	8·5	100
8·0	94·7	5·3	100

Borate Buffer (Holmes, 1943).

$Na_2B_4O_7$, 0·05 M.
Dissolve 19·1 g of $Na_2B_4O_7 \cdot 10H_2O$ in H_2O and dilute to 1000 ml.
(Borax may lose water of crystallisation and should be kept in a stoppered bottle).

H_3BO_3, 0·2 M.
Dissolve 12·4 g of H_3BO_3 in H_2O and dilute to 1000 ml.

Range of pH values which may be obtained by mixing $Na_2B_4O_7$ and H_3BO_3 solutions.

pH	0·05 M $Na_2B_4O_7$ (ml)	0·2 M H_3BO_3 (ml)
7·4	1·0	9·0
7·6	1·5	8·5
7·8	2·0	8·0
8·0	3·0	7·0
8·2	3·5	6·5
8·4	4·5	5·5
8·7	6·0	4·0
9·0	8·0	2·0

C Indicators suitable for colorimetric estimations.
Dissolve 40 mg of the appropriate indicator in about 100 ml of H_2O or alcohol. Adjust the pH (with dilute strong acid or alkali) to near the middle of the range.

Suitable indicators, with the range of pH which each covers, are shown below.

Indicator	pH range
Thymol Blue	1·2–2·8
Bromophenol Blue	3·0–4·6
Methyl Orange	3·1–4·4
Methyl Red	4·4–6·0
Bromocresol Purple	5·2–6·8
Bromothymol Blue	6·0–7·6
*Neutral Red	6·8–8·0
Cresol Red	7·2–8·8
Thymol Blue	8·0–9·6
*Phenolphthalein	8·0–9·8

*Soluble in alcohol but not in H_2O.

Procedure

Follow the maker's instructions. Standardise the instrument with one buffer solution. Use a second buffer solution of pH about 3 units different from the first and check that the instrument gives the correct reading ± 0·1 unit, without any further adjustment of the controls. If it does not there is a fault—usually in the electrode system.

The following notes are intended to supplement the maker's instructions.

(a) Soak new or dried out glass electrodes for several hours in 0·01 M HCl.

(b) Contact between the sample and KCl solution of the salt bridge is established by a small porous fibre, sintered glass disc, or ceramic disc sealed into the lower tip of the KCl holder. If the reference electrode is separate from the glass electrode the KCl holder should be open to the atmosphere and the KCl solution level should be kept above the sample solution level, so that there is a very slow flow of KCl into the sample. The KCl solution in the bridge should not be allowed to dry out or to become unsaturated. It is usual to keep some crystals of KCl in the solution to ensure saturation.

(c) Make sure that air bubbles are not trapped in the ends of either electrode or in the salt bridge.

(d) If the electrodes are separate arrange the tip of the glass electrode slightly above the tip of the reference electrode to minimise danger of breakage. Some electrode stands have no 'stop' or 'collar', and the electrode holder may be slid down too far so that the electrodes are broken. A suitable length of plastic tube slid over the vertical rod of the stand before the holder is slid on will act as a collar and prevent accidental electrode damage.

(e) Changes in temperature and unnecessary contact of the water samples with air should be avoided while making readings. In some circumstances it may be useful to measure the pH both at the *in situ* temperature and at room temperature though in most circumstances a difference of less than 0·1 pH unit will be found.

Avoid contact of the sample with air during storage by filling the bottles completely when sampling. For theoretical aspects see § 9.5.

(f) 'Drifting' of the pH reading may occur if the glass electrode bulb is insufficiently cleaned between samples. Cleaning with cotton wool soaked in detergent may be necessary occasionally. Before measuring poorly buffered samples it is essential to rinse the electrodes with the sample. The pH of unbuffered samples may drift due to absorbtion of CO_2 from the air or to alkali leached from the glass.

(g) Sudden and large changes of reading may be produced by static electrical charges accumulated on clothes worn by the operator. Difficulties from this cause may be reduced by using a special screened glass electrode, or by placing an earthed metal shield round the electrodes.

(h) When not in use the glass electrodes should be immersed in water (or 0·01 M HCl). The electrodes should be kept clean. If the calomel electrode is to be stored for an extended period the protecting cap should be replaced on the end.

(i) When making measurements on samples of pH 9·5 or higher, readings should be corrected for alkali-ion error according to the correction sheet supplied with thc meter. It is better to use a special glass electrode designed for this type of work.

(j) The response of old glass electrodes may become slow. They may be reactivated by dipping in a 20% solution of NH_4HF_2 for a few minutes followed by dipping in 12 M HCl (s.g. 1·19). The accuracy of such reactivated electrodes must be checked carefully.

LEVEL III

3.2.5 Electrometric, with a pH meter and recorder

pH measurements can easily be automated; the pH meter output is fed to a recording device. Battery operated apparatus may be bought.

3.3 OXIDATION-REDUCTION POTENTIAL

Most pH meters are designed so that they may be used to measure small potential differences directly.

Samples must be taken with the same precautions as for pH. Connect a calomel electrode and salt bridge to the reference terminal and a Pt-electrode to the indicator electrode terminal of the pH meter, and switch the range to 'millivolts'. The polarity may be wrong, in which case move the switch to ' — millivolts'. Most scales read from 0 to 1300 or 1400 millivolts. The zero adjuster and temperature compensator are usually automatically disconnected from the circuit when the switch is in a millivolt position.

Using suitable electrodes, for example platinum and calomel, the above procedure can be used to determine oxidation-reduction potentials or to make oxidation-reduction titrations.

It is often difficult to obtain reproducible results. Platinum electrodes may be cleaned either by heating in H_2SO_4 (1 + 1), and leaving to stand in H_2SO_4 (1 + 1) overnight, or by heating to red heat in a gas flame. Calibration of electrodes in absolute terms is very difficult, though relative measurements are fairly simple. (See § 9.11 and Effenberger, 1962).

The continuous measurement of these potentials has also been described by Effenberger (1967). Schindler and Honick (1971) describe an Ag/AgCl-Pt couple for measurements in mud. A salt bridge is not needed.

The interpretation of oxidation-reduction potentials in systems as complex as lake water or sediment is not simple. Golterman (1975) gives some guidance. The pH must be known if the measurement is to have any value at all.

3.4 TOTAL ALKALINITY AND PHENOLPHTHALEIN ALKALINITY

LEVEL II

3.4.1 Acidimetric, indicator end point

Principle

See § 9.7 for further detail.

The amount of $CO_3{}^{2-} + OH^-$ is determined by titration with acid to pH $\simeq$ 8·3, the end point being detected with phenolphthalein.

The amount of HCO_3^- is determined by further titration with acid to an end point pH between 4·2 and 5·4, with methyl orange or mixed indicator as end point indicator.

There are at least 4 quantities commonly reported:

	Quantity	Symbol	Compounds	End point pH
1.	Phenolphthalein alkalinity	PA	$OH^- + CO_3^{2-}$	8·3
2.	Total alkalinity	TA	$OH^- + CO_3^{2-} + HCO_3^-$	4·2 to 5·4
3.	Carbonate alkalinity	CA	$CO_3^{2-} + HCO_3^-$	
4.	Total CO_2	TC	$CO_3^{2-} + HCO_3^- + CO_2$	

CA and TC may be found by calculation from TA, the pH and conductivity of the original water sample.

The exact end point of the TA titration depends on the amount of CO_2 in solution and is therefore related to TC. For precise work therefore the complete titration curve should be obtained potentiometrically. If indicators must be used, an approximate value of TC can be obtained and then a value very close to the true TA end point may be calculated. The titration is taken to this pH as end point.

Total CO_2 (TC) may also be measured directly (see § 3.5). It should be noted that $SiO(OH)_3^-$, $H_3BO_3^-$, NH_4^+, HS^- and some organic anions will be included in the TA estimate. So may colloidal and suspended $CaCO_3$. If the errors thus introduced are not tolerable then a gas chromatographic method, (§ 8.3 or 8.4) in which the sample is acidified and CO_2 measured directly, may be preferred. Gas chromatography may also be necessary if the water has an unusually low concentration of CO_2, as it does for example in regions where the rocks are resistant to weathering.

Reagents

A HCl, 0·100 M (= 100 mmol l^{-1}) or 0·050M (50 mmol l^{-1}). See § 1.3.4.
The concentration to which the standard acid should be diluted depends on the size of the burette and on the order of magnitude of the alkalinity to be determined.

B Phenolphthalein indicator
Dissolve 0·5 g of phenolphthalein in 50 ml of 95% ethanol, and add 50 ml of water. Add dilute (e.g. 0·05 M) CO_2-free NaOH solution dropwise, until the solution turns faintly pink.

C Mixed indicator
Dissolve 0·02 g of methyl red and 0·08 g of bromocresol green in about 100 ml of 95% ethanol. This indicator is suitable over the pH range 4·6–5·2. It is advisable to prepare buffer solutions, (see § 3.2.4) to which is added indicator in the same amount as in an alkalinity titration. These provide standard end points.

D Methyl orange indicator, 0·05%
Dissolve 0·05 g of methyl orange in about 100 ml of water. This indicator is suitable for equivalence points below pH 4·6. Just as for the mixed indicator, it is advisable to prepare a buffer solution to provide standard end points.

E Buffer solution pH 8·3
59·4 ml of 0·05 M $Na_2B_4O_7$ + 40·6 ml of 0·1 M HCl. See § 3.2.4, solutions B, and § 1.3.4.

Procedure

Mix 100 ml of sample with 2 drops of phenolphthalein indicator (B) in a conical titration beaker. If the solution remains colourless PA = 0, and the total alkalinity is determined as described below. If the solution turns red, determine the PA by titrating with standard acid until the colour practically disappears. The very faint colour at the end point is best estimated by comparing with a standard end point in pH = 8·3 buffer (E).

Add 3 drops per 100 ml of sample of either the mixed indicator or methyl orange and determine the TA by continuing the titration to the second equivalence point.

For more precise work read the burette when the first slight tendency to change colour appears. With this approximate value of the TA, and with the initial pH and conductivity of the water, the approximate value of total CO_2 may be found from Tables 3.4 and 3.5. This value for total CO_2 may then be used in Table 3.6 to give the calculated end point pH for the TA titration. This calculated end point pH differs insignificantly from the true end point pH.

Add 3 drops of mixed indicator or methyl orange to 100 ml buffer with the same pH as the calculated end point and continue the titration until the same colour is observed in the sample as in the buffer. Loss of CO_2 must be minimised.

Notes

All reagents should be low in CO_2. This is particularly the case when determining low alkalinities. Dissolved CO_2 may be removed from the dilution water in two ways: (1) reduce the pressure for 10–15 minutes with a water jet pump, or (2) boil for 10–15 minutes and subsequently cool. Reabsorption of CO_2 from the air may be prevented by a soda lime tube.

Calculations

See (§ 9.7 and 9.8) for further explanation.

1. Phenolphthalein alkalinity = $[OH^-] + [\frac{1}{2}CO_3{}^{2-}]$

Let PA = phenolphthalein alkalinity (concentration) [mmol l^{-1}]

$$\text{then} \quad PA = \frac{C_a V_{8\cdot3}}{V_s}$$

where C_a = concentration of acid [mmol l^{-1}]
$V_{8\cdot3}$ = volume of acid used to titrate to pH 8·3 [ml]
V_s = volume of sample [ml]

2. Total alkalinity = $[OH^-] + [\frac{1}{2}CO_3{}^{2-}] + [HCO_3{}^-]$

Let TA = total alkalinity (concentration) [mmol l^{-1}]

$$\text{then} \quad TA = \frac{C_a V_{4-6}}{V_s}$$

where C_a = concentration of acid [mmol l^{-1}]
V_{4-6} = total volume of acid used to titrate to the second end point (pH 4 to 6) [ml]
V_s = volume of sample [ml]

3. Bicarbonate = $[HCO_3{}^-]$

Let x = concentration of $HCO_3{}^-$ [mmol l^{-1}]

$$\text{then } x = TA\text{–}PA = \frac{C_a (V_{8\cdot3} - V_{4-6})}{V_s}$$

Table 3.4. Factors (β) for calculating carbonate alkalinity (CA) from total alkalinity (TA) and from pH and conductivity of the original water sample at 20°C. Units are mmol l^{-1}.

pH_{20}	Conductivity (μS cm^{-1}) 0–40	40–300	300–550
8·8	0	0	0
8·9	1	1	1
9·0	1	1	1
9·1	1	1	1
9·2	1	1	1
9·3	1	1	1
9·4	2	2	2
9·5	2	2	2
9·6	3	3	3
9·7	4	4	4
9·8	4	5	5
9·9	6	6	6
10·0	7	7	7
10·1	9	9	9
10·2	11	11	12
10·3	14	14	15
10·4	17	18	19

4. Carbonate alkalinity = $[\frac{1}{2}CO_3^{\ 2-}] + [HCO_3^-]$

Let CA = carbonate alkalinity (concentration) [mmol l^{-1}]

then CA = TA − 0·01β

where β is a factor with the dimensions of concentration, dependent primarily on $[OH^-]$ and to a smaller extent on ionic strength (estimated from conductivity). The factor β is obtained from Table 3.4 or, with a final error in 0·01 β of less than ±0·01 mmol l^{-1}, by

$$\beta = 0{\cdot}7 \text{ anti-log}_{10}(\text{pH} - 9)$$

Table 3.5. Factors δ for calculating total CO_2 (TC) from carbonate alkalinity (CA), and from pH and conductivity of the original water sample at 20°C. TC=δCA. CA may be obtained from Table 3.4.

pH_{20}	Conductivity ($\mu S\ cm^{-1}$)						
	0–10	10–40	40–110	110–200	200–315	315–430	430–550
6·4	1·96	1·93	1·93	1·90	1·89	1·88	1·86
6·5	1·76	1·74	1·74	1·72	1·71	1·70	1·69
6·6	1·60	1·59	1·59	1·57	1·56	1·55	1·55
6·7	1·48	1·46	1·46	1·45	1·45	1·44	1·43
6·8	1·38	1·37	1·37	1·36	1·35	1·35	1·34
6·9	1·30	1·29	1·29	1·29	1·28	1·28	1·27
7·0	1·24	1·23	1·23	1·23	1·22	1·22	1·22
7·1	1·19	1·19	1·18	1·18	1·18	1·17	1·17
7·2	1·15	1·15	1·15	1·14	1·14	1·14	1·14
7·3	1·12	1·12	1·12	1·11	1·11	1·11	1·11
7·4	1·09	1·09	1·09	1·09	1·09	1·09	1·09
7·5	1·07	1·07	1·07	1·07	1·07	1·07	1·07
7·6	1·06	1·06	1·06	1·06	1·05	1·05	1·05
7·7	1·05	1·05	1·04	1·04	1·04	1·04	1·04
7·8	1·04	1·04	1·03	1·03	1·03	1·03	1·03
7·9	1·03	1·03	1·03	1·03	1·02	1·02	1·02
8·0	1·02	1·02	1·02	1·02	1·02	1·02	1·02
8·1	1·01	1·01	1·01	1·02	1·01	1·01	1·01
8·2	·1·01	1·01	1·01	1·01	1·01	1·01	1·01
8·3	1·00	1·00	1·00	1·00	1·00	1·00	1·00
8·4	1·00	1·00	1·00	1·00	1·00	1·00	1·00
8·5	1·00	·99	·99	·99	·99	·99	·99
8·6	·99	·99	·99	·99	·99	·99	·98
8·7	·99	·98	·98	·98	·98	·98	·98
8·8	·98	·98	·98	·97	·97	·97	·97
8·9	·97	·97	·97	·97	·97	·96	·96
9·0	·96	·96	·96	·96	·96	·95	·95
9·1	·95	·95	·95	·95	·94	·94	·94
9·2	·94	·94	·94	·93	·93	·93	·93
9·3	·93	·93	·92	·92	·92	·91	·91
9·4	·91	·91	·91	·90	·90	·90	·89
9·5	·90	·89	·89	·88	·88	·87	·87
9·6	·88	·87	·87	·86	·85	·85	·85
9·7	·85	·85	·84	·84	·83	·83	·82
9·8	·83	·82	·82	·81	·80	·80	·79
9·9	·80	·79	·79	·78	·77	·77	·76
10·0	·77	·76	·76	·75	·75	·74	·74
10·1	·74	·74	·73	·72	·72	·71	·71
10·2	·72	·71	·70	·70	·69	·69	·68
10·3	·69	·68	·68	·67	·66	·66	·65
10·4	·66	·65	·65	·64	·64	·63	·63

5. 'Total CO_2' = $[\frac{1}{2}CO_3{}^{2-}] + [HCO_3{}^-] + [CO_2]$

Let TC = 'total CO_2' concentration [mmol l^{-1}]

then TC = δ CA

where δ is the factor:

$$\frac{1 + [H^+]/K_1 + K_2/[H^+]}{1 + (2K_2[H^+])}$$ which is a pure number.

This factor may be calculated using:

Temperature °C	0	5	10	15	20	25
10^7 K_1	2·63	3·02	3·47	3·80	4·17	4·47
10^{11} K_2	2·34	2·75	3·24	3·72	4·17	4·68

or obtained from Fig. 9.5 or from Table 3.5.
The factor is to some extent dependent on ionic strength and it is this relationship which is shown in Table 3.5 but for 20°C only.

Table 3.6. End point pH values, at 20°C, for the total alkalinity (TA) titration. Total CO_2 (TC) may be found from Tables 3.4 and 3.5.

Total CO_2 mmol l^{-1}	End point pH	Total CO_2 mmol l^{-1}	End point pH
0·05	5·36	1·2	4·65
0·08	5·25	1·5	4·61
0·10	5·20	2·0	4·54
0·15	5·11	2·5	4·49
0·2	5·05	3·0	4·46
0·3	4·96	4·0	4·39
0·4	4·90	5·0	4·34
0·5	4·85	6·0	4·30
0·6	4·81	7·0	4·27
0·8	4·74	8·0	4·24
1·0	4·69	10·0	4·19

Precision and accuracy

The precision is estimated at 2% at TA = 1 mmol l^{-1} and 2–10% at TA between 1 and 0·1 mmol l^{-1}. Below 0·1 mmol l^{-1} the method is probably unreliable, unless the end point is made sharper by bubbling CO_2-free gas through the solution during the titration. This can best be carried out by putting the sample into a sintered glass filter 15–40 μm mean pore size (e.g. Jena G3) and passing N_2 from below through the filter. Add the sample after the N_2–stream has started and after the filter is wetted. Rinse the sample away with H_2O while the N_2 is still passing through. End point calculation is not valid if this modification is used.

LEVEL II

CONDUCTOMETRIC OR POTENTIOMETRIC TITRATION

Principle

The titration is carried out as described for level I, but a pH meter is used to locate the end point. For dilute solutions a conductance meter is used.

3.4.2 Acidimetric, potentiometric end point

Procedure

Pipette a suitable volume of sample (e.g. 100 ml) into a conical titration beaker placed on a magnetic stirrer. Lower a glass and a calomel electrode, and the capillary tip of a (micro) burette into the sample. Take care that the stirrer cannot damage the electrodes.

For precise work construct a titration curve by plotting the pH values, against the volume of the titrant added, continuing well beyond the end point.

In routine determinations a more rapid method may be used. Take the first end point at pH = 8·3, corresponding to the PA. Determine TA by continuing the titration to the second end point. When the second end point is nearly reached (pH = 5·0–5·5) read the burette. With this approximate value of the alkalinity and the initial pH and conductivity of the water, follow through Tables 3.4, 3.5 and 3.6 (as described in the indicator method, § 3.4.1) to obtain the calculated end point pH. This calculated end point differs insignificantly from the true end point pH. Continue the titration to the calculated end point pH. Use the final reading of the burette for the calculation of the exact TA and total CO_2. Loss of CO_2 must be minimised.

Notes and calculation

See (§ 3.4.1).

Precision and accuracy

The precision and accuracy are estimated at 1% or better at TA > 1 mmol l^{-1} and at 1 to 5% at TA ⩽ 1 mmol l^{-1}. Total CO_2 is most accurately calculated in comparatively soft waters and in waters where the pH lies between 7·0 and 9·2, but an absolute condition is that the pH is measured accurately. The error in total CO_2 arising from an error in the pH measurement of 0·1 pH–unit is:

pH	6·4	7·4	8·4	9·4	10·4
Error in TC [%]	10·8	2·0	0·5	2·6	5·2

For unfamiliar waters it is wise to check the calculation method by some independent method involving the liberation of CO_2, subsequent absorption of the CO_2 in another solvent, and measurement of CO_2 in the second solvent.

Interferences

Precipitation may occur during storage if the sample is supersaturated with $CaCO_3$. This will affect the alkalinity determination. In such cases the alkalinity titration should be performed as soon as possible after sampling.

3.4.3 Acidimetric, conductometric end point

Procedure

For waters of low alkalinity the same procedure as (§ 3.4.2) can be used with a conductance meter as end point detector instead of a pH meter.

Most devices for conductance measurement can be used, as it is sufficient to determine relative units. Meters with a titration attachment are preferable. A cell constant of 1·0 cm^{-1} is suitable. The Pt–plates should be covered by Pt–black (see § 3.1.1).

Frequencies from 60 Hz up to 1000 Hz and potential differences of 5–30 V are suitable for the analysis.

Conductance should be plotted against volume of titrant as described in § 9.1 and § 9.7, Fig. 9.3B.

LEVEL III

3.4.4 Acidimetric, potentiometric end point with recorder

The same method as in level II is used but with a pH meter with recorder connected to a motor-driven piston burette or automatic titrator. As the pH of the end point depends on the water to be titrated the recording type of apparatus is preferable to the pre-set end point type.

LEVEL II or III

3.4.5 Acidimetric, 'Gran titration'

Six potentiometric measurements are made and a novel form of graphic plot of antilog potential against volume of titrant is used for calculation. The whole analysis takes 5–10 minutes. Practical details are given by Talling (1973) in whose hands the method had precision of about 0·001 mmol l^{-1}. This method seems particularly promising for the determination of small changes in concentration. It was proposed by Gran (1952) and developed by Edmond (1970).

3.5 TOTAL CO_2

LEVEL II

Principle

Free CO_2 and CO_3^{2-} are both converted to HCO_3^-; free CO_2 by titration with (carbonate-free) NaOH; CO_3^{2-} by titration with HCl. Both titrations have end point pH = 8·3. The HCO_3^- concentration is then determined with HCl.

Reagents

A HCl 0·1 M or 0·05 M See § 3.4.1.

B NaOH 0·1 M or 0·05 M, carbonate free

Dissolve 50 g of NaOH (A.R.) in 50 ml H_2O in a Pyrex flask. Filter this concentrated caustic solution through a sintered glass filter (excluding air) to remove the Na_2CO_3. The strong carbonate-free NaOH solution can be kept in thick walled polythene or polypropylene bottles. The solution is about 19 M.

Dilute with CO_2-free (boiled) H_2O to the desired strength. Exclude air. Standardise against standard acid.

If the standard solutions are transferred quickly to a burette, contamination by CO_2 may be disregarded. The burette must be protected by a soda lime tube. It is better however to connect the burette directly to the stock bottle.

NaOH may be tested for carbonate by mixing 20 ml with 1 ml of 0·5 M $BaCl_2$ out of contact with air. No turbidity should appear.

3.5.1 Acidimetric, indicator end point

Procedure

Add several drops of phenolphthalein indicator (§ 3.4.1) to 100 ml of the sample. Then titrate with 0·05 M NaOH until the solution has a faint pink colour; if the solution turns pink when the indicator is added, titrate with 0·05 M HCl instead. Then add the mixed indicator (§ 3.4.1, solution C) and titrate with 0·05 M HCl solution. The end point may be sharpened by bubbling N_2 through the liquid to remove CO_2.

Precision and accuracy

As for § 3.4.1.

Interferences

The estimation of bicarbonate in waters heavily contaminated with organic matter or in distinctly saline waters is best done by distilling off the CO_2 from an acidified sample, and estimating the amount of this CO_2. This may be conveniently done in a Warburg apparatus, or gas analysis apparatus.

Calculation

Let x = concentration of CO_2 in the water sample [mmol l^{-1}]

then $$x = \frac{C_a (V_2 - V_1)}{V_s}$$

where C_a = concentration of HCl [mmol l^{-1}]
$V_2 - V_1$ = difference in volume of HCl at the second and first equivalence points [ml]
V_s = volume of sample [ml]

3.5.2 Acidimetric, potentiometric end point

Same as § 3.5.1 but with potentiometric end point (see § 3.4.2 or § 3.4.4).

LEVEL III

3.5.3 Gas chromatographic

See § 8.3 or § 8.4.

3.6 ACIDITY

LEVEL II

Principle

The sample is titrated with dilute $Ba(OH)_2$ to pH 8·6. Many water samples contain no titratable acid.

Reagents

A $\frac{1}{2}Ba(OH)_2$, 0·01 M = 10 mmol l^{-1}

Dissolve 3·2 g of $Ba(OH)_2 \cdot 8H_2O$ in 2 litres of boiled H_2O. Store the solution in a container which is connected directly to a burette. Exclude CO_2 from both with a soda-lime tube. Allow any $BaCO_3$ precipitate to settle before transferring the solution to the burette.

Standardise with HCl.

B Mixed indicator.

Mix 10 ml of 0·1% Thymol Blue (in 50% alcohol) with 30 ml of 0·1% phenolphthalein (in 50% alcohol). At pH 8·6 the colour changes from yellow (acid) to red violet (alkaline).

Procedure

Put a suitable accurately measured volume, for example 50·0 ml, of sample in a titration flask. Add one drop of indicator. Titrate with $Ba(OH)_2$, solution A, to the end point.

The end point may be detected with a pH meter rather than by indicator.

Calculation

Let x = acidity of the water sample [mmol l^{-1}]

then $$x = \frac{C_b V_b}{V_s}$$

where C_b = concentration of $Ba(OH)_2$ [mmol l^{-1}]
V_b = volume of $Ba(OH)_2$ [ml]
V_s = volume of sample [ml]

CHAPTER 4

MAJOR CONSTITUENTS*

CALCIUM, MAGNESIUM, SODIUM, POTASSIUM, IRON, CHLORIDE, SULPHATE, SULPHIDE, CARBONATE AND BICARBONATE

4.1 CALCIUM

LEVELS I AND II

4.1.1. Colorimetric

Principle
Calcium forms a coloured complex with glyoxalbis-(2-hydroxyanil) = di-(o-hydroxyphenylimino)ethane (Kerr, 1960).

Reagents

A_1 Standard Ca^{2+} solution, 400 $\mu g\ ml^{-1}$
Dissolve 1·00 g of $CaCO_3$ (dry, A.R.) in 25 ml of 1 M HCl and dilute to 1000 ml with H_2O.

A_2 Standard Ca^{2+} solution, 4 $\mu g\ ml^{-1}$ (=0·2 mmol l^{-1} $\frac{1}{2}Ca^{2+}$)
Dilute 5·0 ml of stock solution A_1 to 500 ml.

B Buffer solution, pH=12·6
Dissolve 10 g of NaOH and 10 g of $Na_2B_4O_7$ in 1 litre of H_2O.

C Glyoxalbis-(2-hydroxyanil), 0·5% in methanol
This substance may be either purchased or prepared thus:
Dissolve 4·4 g of freshly sublimed o-aminophenol in 1 litre of H_2O at 80°C; add 3·5 ml of glyoxal solution (30% w/w); maintain the temperature at 80°C for 30 minutes, cool, and allow to stand for 12 hours in a refrigerator. Filter off the precipitate, wash with water, and recrystallize from methanol.
Dissolve 0·5 g of glyoxalbis-(2-hydroxyanil) in about 100 ml of methanol.

D Ethanol-n-butanol solvent mixture
Mix equal volumes of ethanol and n-butanol.

Procedure
Put 20·0 ml of sample containing not more than 80 μg of Ca^{2+} (= 4 μmol $\frac{1}{2}Ca^{2+}$), or an aliquot diluted to 20·0 ml, into a large test tube. Add in succession 2·0 ml of buffer (B), 1·0 ml of glyoxalbis-(2-hydroxyanil) reagent (C) and 20·0 ml of solvent mixture (D), mixing after each addition. Measure the absorbance of the reagent

*MAJOR CONSTITUENTS is a term of convenience. Included are most of the substances contributing 1% or more to the total of inorganic solutes.

blank, standards and samples in suitable cells at a wavelength as close as possible to 520 nm. The measurement should be made not less than 25 minutes after mixing. The colour is stable for at least 35 minutes after this. Prepare a calibration curve covering the range 0–80 μg of Ca^{2+}(=0–4·0 μmol $\frac{1}{2}Ca^{2+}$).

Interferences
According to Kerr (1960), up to 50 ppm of Mg^{2+} and Al^{3+} do not interfere; Ba^{2+} and Sr^{2+} do interfere above 4 ppm; Fe^{2+} has no effect if precipitation is prevented by adding cyanide; SO_4^{2-}, Cl^-, and NO_3^- are without effect up to 500 ppm; Cu^{2+} above about 0·1 ppm appears to delay colour formation but this interference can be suppressed by adding a few drops of a 2% solution of sodium diethyldithiocarbamate before adding the reagents.

Sensitivity
0·1 mg l^{-1}=0·005 mmol l^{-1} $\frac{1}{2}Ca^{2+}$

Precision and accuracy
Estimated at 5%.

4.2 MAGNESIUM

LEVEL I

4.2.1. Colorimetric

Principle
Magnesium forms a coloured complex with the dye Brilliant Yellow (Taras, 1948).

Reagents
Solution A_2 is alternative to A_1.

A_1 Standard Mg^{2+} solution, 1 mg ml^{-1}.
Dissolve 10·135 g $MgSO_4.7H_2O$ in H_2O and dilute to 1000 ml.

A_2 Standard Mg^{2+} solution, 0·1 mmol ml^{-1} $\frac{1}{2}$ Mg^{2+}
Dissolve 12·315 g $MgSO_4.7H_2O$ in H_2O and dilute to 1000 ml.

B H_2SO_4, 0·05 M
Add 5 ml of H_2SO_4 (1+1) (see § 1.3.4) to 1 litre of H_2O.

C NaOH, 5 M
Dilute 10 M NaOH (see § 1.3.4) 0·5 litre → 1·0 litre.

D Ca^{2+}-Al^{3+} solution
Dissolve 1·5 g of $Al_2(SO_4)_3.18\ H_2O$ in 500 ml of H_2O. Add 30 g of $CaCO_3$ and 40 ml of HNO_3 (s.g. 1·42), taking care to avoid excessive effervescence. Make up to 1 litre.

E Brilliant Yellow solution
Dissolve 10 mg of Brilliant Yellow (B.D.H.) in 0·1 litre of H_2O. Prepare a fresh solution every 2 or 3 days.

F Starch or Tylose® H_{10} solution 2%
Dissolve 1 g of soluble starch in 50 ml of H_2O. Heat to aid solution. The ethyl cellulose derivative Tylose® H_{10} is much superior to starch.

Procedure
Put 50·0 ml of sample containing not more than 0·5 mg of Mg^{2+} (about 0·05 mmol $\frac{1}{2}$ Mg^{2+}), or an aliquot diluted to 50·0 ml, into a 100 ml volumetric flask. Add in succession 1 ml of 0·05 M H_2SO_4 (B), 1 ml of Ca^{2+}-Al^{3+} solution (D), 5 ml of starch solution (F) and 4 ml of NaOH (C). Rinse the neck of the flask with H_2O and add 10 ml reagent solution (E). Mix well and make up to 100 ml with H_2O. Measure the absorbance of the reagent blank, standards and samples in suitable cells at a wavelength as close as possible to 540 nm after 15 to 20 minutes. Prepare a calibration curve using known amounts of Mg^{2+} covering the range 1·0–10·0 mg l^{-1} (0·1 to 1·0 mmol l^{-1} $\frac{1}{2}$ Mg^{2+}). The calibration curve is not linear.

Interferences
Normally interference would be expected from varying concentrations of Ca^{2+} and Al^{3+} in the sample, but these variations are largely avoided by the addition of the relatively large quantities of these elements.

Precision, accuracy and sensitivity
Precision and accuracy are unknown. The sensitivity is about 0·5 mg l^{-1} (about 0·04 mmol l^{-1} $\frac{1}{2}$ Mg^{2+}).

4.3 CALCIUM AND MAGNESIUM

LEVEL II

Volumetric with EDTA

Principle

METHOD 4.3.1
EDTA* is added to a solution of Ca^{2+} and Calcon.† The last two form a pink coloured complex. As EDTA forms a stronger complex with Ca^{2+} than does Calcon the solution changes from pink to the blue colour of Calcon itself. Mg^{2+} is first precipitated by adding NaOH, so that Ca^{2+} alone is determined. The total of Ca^{2+} plus Mg^{2+} is determined according to the same principle in a second sample with Eriochrome Black T‡ at pH = 10.

METHOD 4.3.2
The same principle as in method 1, but with glyoxalbis-(2-hydroxyanil) as indicator, again at pH = 12·6. After the Ca^{2+} determination, acid is added to dissolve the Mg precipitate and to destroy the indicator. Mg^{2+} is then determined according to method 4.3.1.

*Na_2EDTA = disodium dihydrogen ethylenediaminetetra-acetate. Flaschka (1964) discusses EDTA titrations.

†Calcon = Solochrome Dark Blue = Eriochrome Blue Blank R = CI-Mordant Black 17.

‡Eriochrome Black T = Solochrome Black T = CI-Mordant Black 11. (Probably better, but sometimes difficult to obtain is Solochrome Black 6B = Eriochrome Blue-Black B = CI-201).

4.3.1. With Calcon and Eriochrome Black T end point

Reagents

A $\frac{1}{2}$ Na_2 EDTA, 0·020 M = 20 mmol l^{-1}
Dilute 0·200 M $\frac{1}{2}$ Na_2EDTA (see § 1.3.4) 100 ml → 1000 ml. 1 ml is equivalent to 0·4 mg Ca^{2+} or 0·02 mmol $\frac{1}{2}$ Ca^{2+} .The solution may be stored in Pyrex glass vessels or in thick walled polythene bottles, through which evaporation is negligible.

B_1 Ca^{2+} + Mg^{2+} solution
Mix 25·0 ml of standard Ca^{2+} solution A_1 of § 4.1.1 and 10·0 ml of standard Mg^{2+} solution (A_1) of § 4.2.1 and dilute to 1000 ml; 1 ml of this solution contains 10 μg of Ca^{2+} and 10 μg of Mg^{2+}.

B_2
Mix 25·0 ml of standard Ca^{2+} solution A_1 of § 4.1.1 and 10·0 ml of standard Mg^{2+} solution A_2 of § 4.2.1 and dilute to 1000 ml; 1 ml of this solution contains 0·5 μmol ml^{-1} $\frac{1}{2}$ Ca^{2+} and 1·0 μmol ml^{-1} $\frac{1}{2}$ Mg^{2+}.

Solution B_2 is alternative to B_1.

C_1 Borax-buffer for Ca^{2+} + Mg^{2+}, pH 12
Dissolve 8 g of $Na_2B_4O_7.10H_2O$ in 160 ml of H_2O. Add to this solution 2 g of NaOH + 1 g of $Na_2S.9H_2O$ previously dissolved together in 20 ml of H_2O, and dilute the mixed solutions to 200 ml.

C_2
Dilute C_1 10 times with H_2O directly before use.

D NaOH, 0·1 M
Dissolve 4 g of NaOH in 1 litre of H_2O.

E HCl, 1 M See § 1.3.4

F_1 Calcon indicator*
Grind together 0·20 g of powdered Calcon and 100 g of solid NaCl (A.R.) and keep in a stoppered bottle. Keep dry.

F_2 "Glyoxal" indicator solution
Dissolve 0·03 g of glyoxalbis-(2-hydroxyanil) = di-(o-hydroxyphenylimino)ethane in about 100 ml of methanol.

G Eriochrome Black T indicator*
Grind together 0·4 g of Eriochrome Black T and 100 g of solid NaCl (A.R.) and keep in a stoppered bottle. Keep dry.

Procedure

Calcium
Prepare a standard end point. Put 40·0 ml of the Ca^{2+} + Mg^{2+} solution (B_1 or B_2) in a 100 ml flask, add 5 ml of 0·1 M NaOH (D) and 100–200 mg of indicator mixture (F_1) (a 'constant spoonful'). From a 2 or 5 ml microburette run in slowly

*See footnotes, p. 69.

1·00 ml of EDTA solution (A). The pink colour will change to blue, which is the end point colour. Repeat the procedure with the unknown sample adding the EDTA solution slowly until the colour matches that of the standard end point.

Calcium plus Magnesium

Prepare a standard end point. Put 40·0 ml of the Ca^{2+} + Mg^{2+} solution (B_1) in a 100 ml flask, add 1 ml of diluted buffer solution (C_2) and about 100 mg of Eriochrome Black T indicator (G) (a 'constant spoonful'). Heat at 70°C and run in slowly 2·645 ml of EDTA solution (A) from a piston microburette; the wine-red colour changes to blue. (The total of 2·645 ml is made up of 1·000 ml equivalent to the Ca^{2+} + 1·645 ml equivalent to the Mg^{2+}.) Alternatively use 40·0 ml of the Ca^{2+} + Mg^{2+} solution B_2 and 3·000 ml of EDTA solution (A). Mix an aliquot of the sample containing not more than 0·4 mg of Ca^{2+} (0·02 mmol $\frac{1}{2}$ Ca^{2+}) and not more than 0·4 mg of Mg^{2+} (about 0·03 mmol $\frac{1}{2}$ Mg^{2+}) with 1 ml of the diluted buffer (C_2) in a 100 ml flask. (If the water contains more than 0·3 mmol l^{-1} HCO_3^- add first an equivalent amount of dilute HCl, and boil to remove the CO_2. This prevents $CaCO_3$ precipitation.) Add the same amount of indicator (G) and titrate at 70°C as described above to the colour of the standard end point.

Interferences

Relatively high concentrations of Mn^{2+} lead to an unsatisfactory end point. A remedy is described by Cheng, Melsted and Bray (1953).

Calculation

Let x = concentration of $\frac{1}{2}$ Ca^{2+} in the water sample [mmol l^{-1}]

then $$x = \frac{C_a V_{a1}}{V_s}$$

where C_a = concentration of $\frac{1}{2}$ Na_2 EDTA, reagent A [mmol l^{-1}]
V_{a1} = volume of $\frac{1}{2}$ Na_2EDTA titrant to the first end point [ml]
V_s = volume of water sample [ml]

Let y = concentration of $\frac{1}{2}$ Ca^{2+} in the water sample [mg l^{-1}]

then $$y = 20{\cdot}0\,x = \frac{20{\cdot}0\,C_a V_{a1}}{V_s}$$

Let u = concentration of $\frac{1}{2}$ Mg^{2+} in the water sample [mmol l^{-1}]

then $$u = \frac{C_a (V_{a2} - V_{a1})}{V_s}$$

where V_{a2} = volume of $\frac{1}{2}$ Na_2 EDTA titrant to the second end point [ml]

Let w = concentration of $\frac{1}{2}$ Mg^{2+} in the water sample [mg l^{-1}]

then $$w = 12{\cdot}15\,u = \frac{12{\cdot}15\,C_a (V_{a2} - V_{a1})}{V_s}$$

Let z = concentration of $\frac{1}{2}$ Ca^{2+} + $\frac{1}{2}$ Mg^{2+} in the water sample [mmol l^{-1}]

Then $$z = \frac{C_a V_{a2}}{V_s}$$

Note
If Mg^{2+} is present in much smaller concentrations than Ca^{2+} the EDTA difference method is relatively inaccurate. In these cases more accurate results for Mg^{2+} will be obtained with the colorimetric method § 4.2.1 or method § 4.3.2.

4.3.2. With glyoxalbis-(2-hydroxyanil) and Eriochrome Black T end points

Reagents See (§ 4.3.1)

Procedure

Calcium
Prepare a standard Ca^{2+} end point as in § 4.3.1, but with 3 ml indicator F_2; the colour will change from orange red to lemon yellow.

Titrate an aliquot of the sample, containing not more than 0·4 mg of Ca^{2+} (= 0·02 mmol $\frac{1}{2}$ Ca^{2+}), in the same way.

Magnesium
After the calcium titration, acidify the same samples with 1 M HCl (E) to pH 4·0. This destroys the colour of the "glyoxal indicator", and dissolves the $Mg(OH)_2$. Then add 1 ml of buffer solution (C_2) and indicator Eriochrome Black T (G). Determine the end points using a standard end point as described in method § 4.3.1.

Interferences
Al^{3+} interferes in the Mg^{2+} determination when 1 mg l^{-1} or more is present.

Calculation
See § 4.3.1.

4.3.3. Calcium and Magnesium, potentiometric end point

Principle
Ca^{2+} is titrated potentiometrically with EDTA solution using a Hg–Hg EDTA indicator electrode and a calomel reference electrode. The titration is carried out above pH = 12 at which Mg^{2+} precipitates. After acidification (to dissolve the $Mg(OH)_2$) and subsequent adjustment of the pH to 10·0, the Mg^{2+} can be titrated.

Apparatus
Recording potentiometric titrator, with J-type Hg-electrode (see Figure 4.1), glass and calomel electrodes.

Reagents

A $\frac{1}{2}$ Na_2 EDTA, 0·050 M = 50 mmol l^{-1}
Dilute 0·200 M $\frac{1}{2}$ Na_2 EDTA (see § 1.3.4) 250 ml → 1000 ml.

B NaOH, 10 M See § 1.3.4

C HCl, 1 M See § 1.3.4

D Diethylamine buffer (pH = 11·5)
Dissolve 5 g of diethylamine and 1 g of NH_4Cl in about 100 ml of H_2O.

E Hg-EDTA solution, 0·002 M (highly poisonous)
Prepare 0·025 M $Hg(NO_3)_2$ by dissolving 0·850 g $Hg(NO_3)_2.H_2O$ in 100 ml H_2O Mix equal volumes of this solution and of solution (A). Before use, dilute 20 ml of this mixture to 250 ml with H_2O.

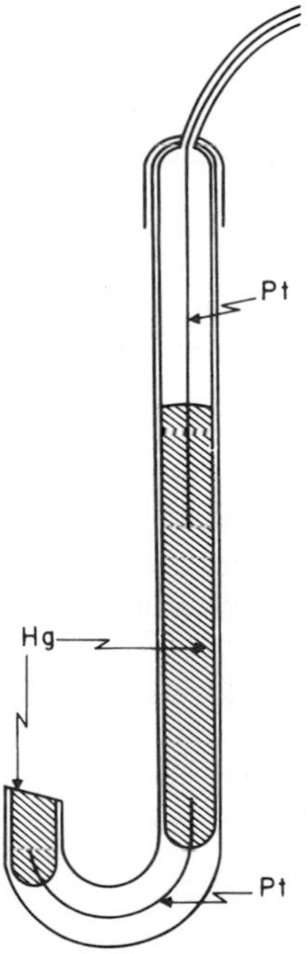

Figure 4.1. J type mercury electrode.

Procedure
Put 50 ml of the sample containing not more than 1·0 mg of Ca^{2+} and 1·0 mg of Mg^{2+} (0·05 mmol $\frac{1}{2}$ Ca^{2+} and about 0·1 mmol $\frac{1}{2}$ Mg^{2+}), or an aliquot diluted to 50 ml, in a 100 ml beaker with a slight excess of HCl and boil to remove the CO_2. (The samples used for determining alkalinity, procedure § 3.4.2, can be used if no coloured indicators have been added). If the volume is too small to cover the electrodes, add H_2O which has been acidified with 5 ml 0·1 M HCl per litre and boiled to remove CO_2. Add 0·4 ml of the Hg-EDTA solution (E), adjust to pH = 12 with NaOH (B) and titrate with EDTA solution (A) from a suitable microburette to a point beyond the end point of the Ca^{2+} titration.

Adjust the pH to 4·0 with HCl (C) and leave until the $Mg(OH)_2$ is dissolved (about 1 minute). Add diethylamine buffer (D) sufficient to make the pH = 10·9 and continue the titration with EDTA to the end point. Prepare a standard end point with solution B_1 or B_2 of § 4.3.1.

Interferences
Fe^{2+}, when more than 1·0 mg l^{-1} is present, must be masked with o-phenanthroline, see § 4.5.1. When more than 28 mg l^{-1} is present it must be removed.

Cu^{2+}, Mn^{2+}, and Zn^{2+} at concentrations up to 10 mg l^{-1} do not interfere.

Calculation
See § 4.3.1.

LEVEL III

4.3.4. Calcium and Magnesium, by atomic absorption flame spectrophotometry

Principle
See § 1.4.3.

Instruments suitable for limnological work are expensive. Cheaper instruments may not be sufficiently sensitive. The method is dependent on the nature and characteristics of the flame, so no specific details can be given here. The manufacturer's instructions should be followed.

Interferences
The presence of interference may be discovered by the method of standard addition (§ 1.2.3).

In the air-acetylene flame Na^+ and K^+ cause some enhancement of absorption by both Ca^{2+} and Mg^{2+}. More serious is the suppression of dissociation by SO_4^{2-}, PO_4^{3-}, 'SiO_2' and Al^{3+}. The interferences may be largely avoided by using the hotter nitrous oxide-acetylene flame, but ionisation of Ca^{2+} and Mg^{2+} is then more serious and must be corrected by adding 2000–5000 mg l^{-1} K^+. Alternatively the depression in the air-acetylene flame may be corrected by adding 5000 mg l^{-1} Sr^{2+} or 10000 mg l^{-1} La^{3+}. The concentrations needed depend on specific conditions and apparatus. Dinnin (1960) gives an example.

Reagents (for suppressing interference)
Because relatively huge amounts of lanthanum or strontium must be added, tiny traces of impurities become important. Special grades of salts with guaranteed purity and suitable for AAFS work can and should be obtained. They are expensive. Solutions A_1, A_2 and A_3 are alternatives.

A_1 Lanthanum solution, 5% w/v
Wet 60 g of La_2O_3 with 100 ml of H_2O. Add *slowly* and whilst *stirring* 250 ml concentrated HCl (CAUTION: the reaction is violent). Allow to cool and make up to 1 litre with H_2O. Solutions of $LaCl_3$ specially prepared for AAFS work may be bought. They are usually 5% or 10% w/v.

A_2 Lanthanum solution, 5% w/v
In 0·5 litre H_2O dissolve 156 g $La(NO_3)_3.6H_2O$. Make up to 1 litre with H_2O.

A_3 Strontium solution, 10% w/v
In 0·5 litre of H_2O dissolve 305 g $SrCl_2.6H_2O$. Make up to 1 litre with H_2O.

Procedure
For air-acetylene flame:
To 4·00 ml sample add 1·00 ml A_1, A_2, or A_3.

(Adjust the proportions if commercial solutions of different concentration are used). *Both* volumes must be measured accurately *and* precisely (§ 1.2.1).

Element	Wavelength (nm)	*Relative absorbance	Flame
Ca	422·7	1·0	reducing
	239·9	200	
Mg	285·2	1·0	oxidising
	202·5	50	

*The actual absorbance (with optimum adjustment of controls) depends on the machine. For the more sensitive wavelengths the useful working range would typically be 1 to 40 mg l^{-1} Ca^{2+} (0·05 to 2·0 mmol l^{-1} $\frac{1}{2}$ Ca^{2+}) and 1 to 4mg l^{-1} Mg^{2+} (0·01 to 0·04 mmol l^{-1} $\frac{1}{2}$ Mg^{2+}).

It may be preferable to use a nitrous oxide-acetylene flame. Consult the manufacturer's handbook.

If calcium-containing particles are present the sample must be digested before analysis. Procedures in § 6.0.2 may be suitable, but the persulphate method is not.

Further information may be found in Fassel and Mosotti (1965), Herrmann (1965), Walsh (1965), Elwell and Gidley (1966), Reynolds and Aldous (1969), Price (1972), Christian and Feldmann (1970).

4.4 SODIUM AND POTASSIUM

LEVEL II AND III

4.4.1. Flame emission spectrophotometric

Principle

Measurement of the light emitted by Na^+ and K^+ when these elements are excited in a flame.

(K: 769 nm; Na: 589 nm).

Procedure

Follow the instructions provided by the manufacturer. The samples must be free of solids.

Prepare a calibration curve. This is often appreciably non-linear. In the case of determination of Na^+, dilute the sample until the Na^+ content is lower than 5 mg l^{-1} (about 0·25 mmol l^{-1} Na^+).

Interferences

The nature and extent of interferences depend very much on the characteristics of the flame, and hence on the particular instrument used. The method of standard addition (§ 1.2.3) should be used.

No detailed guidance can therefore be given, but it may be noted that in some cases mutual interference of Na^+ and K^+ may be serious. In this case make a K^+ calibration curve using K^+ solutions containing the same concentration of Na^+ as is present in the sample and reciprocally for Na^+. Anion interferences (particularly PO_4^{3-} and SO_4^{2-}) may also be troublesome.

Note

Collect and store samples in polyethylene or polypropylene bottles only.

4.5 IRON See also (§ 6.4)

LEVEL I AND II

4.5.1. Colorimetric with o-phenanthroline

Principle

Iron after reduction to Fe^{2+} reacts with o-phenanthroline to form a red compound (see e.g. O'Connor *et al.*, 1965, Nicolson, 1966). Total iron can be estimated after digestion and reduction. Iron may be present in water in 'ionic' form (e.g. as Fe^{2+} or Fe^{3+}), as dissolved complexes (e.g. as humic compounds), colloidal, and as insoluble suspended matter. Fe^{2+} will be oxidised as soon as O_2 is allowed to enter and the Fe^{3+} will normally precipitate, except in acid waters.

For further remarks, see § 6.0 and § 6.4

Reagents

A Standard Fe^{2+} and Fe^{3+} solutions. See § 6.4

B_1 H_2SO_4 [1 + 1]
See § 1.3.4. Use iron free grade.

B_2 H_2SO_4, 0·005 M
Dilute B_1 2000 times (0·5 ml → 1 litre).

C NH_4OH, 4 M
Dilute NH_4OH (s.g.0·91) (0·3 litre → 1·0 litre).

D $NH_2OH.HCl$, 1·0%
Dissolve 1·0 g of $NH_2OH.HCl$ in about 100 ml of 0·005 M H_2SO_4 (B_2). If traces of iron must be removed see § 6.4, reagent (F).

E o-Phenanthroline solution, 0·5%
Dissolve 0·5 g of o-phenanthroline in about 100 ml of 0·005 M H_2SO_4 (B_2).

Procedure

Fe^{2+}
(in ionic form)
Add 1 ml of reagents (B_2) and (E) to 25 ml of sample. The pH should now be between 2 and 3. Measure the absorbance of the reagent blank, standards and samples in suitable cells at a wavelength between 490 and 510 nm and follow the change of the absorbance with time. If only ionic Fe^{2+} is present, the maximum is reached in 30 minutes, but if other forms of Fe are present they may also react, but more slowly, and the absorbance will continue to increase for some hours. Whether or not it is practicable to estimate the amount of Fe^{2+} will depend on the occurrence either of the maximum after about 30 minutes, or of a distinct break in the curve of absorbance against time.

Make a calibration curve covering the range of 50 $\mu g\ l^{-1}$ to 5 $mg\ l^{-1}$ of Fe.

Fe^{3+} + Fe^{2+}
(in ionic form)
Add 1 ml of reagent (D) (instead of B_2) to 25 ml of sample. Proceed as for Fe^{2+}. Follow the change of absorbance with time.

Total Fe

Digest 25 ml, or a measured larger volume, of sample with 2–4 ml H_2SO_4 (B_1) and, if the destruction of the colour takes too long a time, with $K_2S_2O_8$. Add 10 ml of H_2O and wait until all Fe is dissolved. (A rusty circle on the glass will sometimes take 24 hours to dissolve.) Adjust the pH to 2–3 with NH_4OH (C) and adjust the volume to 50 ml. Add 1 ml of reagents (D) and (E) to 25 ml and proceed as for Fe^{2+}.

Interferences

Other complex-forming substances such as phosphate and humic acids keep the iron from forming the iron-phenanthroline complex. The colour will develop slowly in such cases (ionic Fe^{2+} takes about 30 min.). The samples can, however, easily be left overnight. In some waters acidification to pH = 2 to 2·5 is sufficient to render the Fe reactive. Boiling the sample will help, but will reduce Fe^{3+} to Fe^{2+}. All acids, including the reagents (D) and (E), may convert complex iron to the ionic state.

Self colour of the water sample may be corrected by measuring the absorbance of a sample blank, made from sample +1 ml solution (B_2)+1 ml solution (D).

Precision, accuracy, sensitivity

The precision is estimated at 1–2%. The accuracy depends on the interfering substances. The sensitivity is 10–50 $\mu g\ l^{-1}$ of Fe, depending on the colour of the water. For a method with greater sensitivity see § 6.4.

LEVEL III

4.5.2. Iron by atomic absorption flame spectrophotometry (AAFS)

Iron at levels of about 1 mg l^{-1} may be estimated by AAFS. The sample may be concentrated by extracting the (batho) phenanthroline complex with an organic extractant.

Wavelength: 248·3 nm. Oxidising air-acetylene flame.

4.5.3. Iron by a polarographic method

The determination of ferrous and manganous ions in anaerobic hypolimnetic waters has been described by Davison (1976 and 1978), who compared differential pulse and D.C. sampled polarography of both ions.

4.6 CHLORIDE

LEVEL I

4.6.1. Volumetric; with indicator

Principle

The chloride is titrated with $Hg(NO_3)_2$ at a pH of 3·1. (Bromphenol blue). $HgCl_2$ is formed which is slightly dissociated. At the end point the excess Hg^{2+} produces a violet colour with diphenylcarbazone.

Reagents

A Standard NaCl, 0·0100 M
Dissolve 5·845 g of NaCl (dry A.R.) in H_2O and make up to 1000 ml.
Dilute by 10 times (100 ml → 1000 ml).

B $Hg(NO_3)_2$, 0·01 M = 10 mmol l^{-1}
This reagent is highly poisonous.

Dissolve 3·4 g of $Hg(NO_3)_2.H_2O$ in 800 ml of H_2O to which 20 ml of 2 M HNO_3 has been added. Dilute to 1000 ml. Standardize against NaCl as described in the procedure.

C HNO_3, 0·2 M
Dilute 13 ml of HNO_3 (A.R., s.g. 1.42) to 1 litre.

D Diphenylcarbazone-bromphenol blue mixed indicator solution
Dissolve 0·5 g of diphenylcarbazone and 0·05 g of bromphenol blue in about 100 ml of 95% ethyl alcohol or methylated spirit. The solution is stable if kept in a brown bottle.

Procedure

Mix 100 ml of sample containing not more than 10 mg of Cl^- (about 0·3 mmol l^{-1} Cl^-), or an aliquot diluted to 100 ml, with 10 drops (0·5 ml) of indicator solution D. (If more than 10 mg l^{-1} SO_3^{2-} is present add 3 drops (0·15 ml) of H_2O_2.) Add 0·2 M HNO_3 (C) dropwise until the solution becomes yellow (pH = 3·6). Add 5 drops (0·25 ml) more of 0·2 M HNO_3 (C).

Titrate with $Hg(NO_3)_2$ (B) to the point where the first tinge of blue-purple appears which does not disappear on shaking. Prior warning that the end point is near is given when the colour changes to orange.

Prepare a blank to identify the end point colour.

Interferences

I^-, Br^-, CNS^- and CN^- are measured as equivalents of Cl^-. The ions CrO_4^{2-}, Fe^{3+}, Mn^{2+}, Zn^{2+}, and Cu^{2+} react with the diphenylcarbazone and must be removed if present in concentrations greater than 10 mg l^{-1}.

Calculation

Let x = concentration of Cl^- in the water sample [mmol l^{-1}]

then
$$x = \frac{C_b V_b}{V_s}$$

where C_b = concentration of reagent B [mmol l^{-1}]
V_b = volume of reagent B [ml]
V_s = volume of sample [ml]

Let y = concentration of Cl^- in the water sample [mg l^{-1}]

then
$$y = 35{\cdot}5\,x = \frac{35{\cdot}5\,C_b V_b}{V_s}$$

Precision and accuracy

Estimated at:

$0{\cdot}5$ mg l^{-1} from 0 to 50 mg l^{-1};
$1{\cdot}0$ mg l^{-1} from 50 to 100 mg l^{-1}

or about

$0{\cdot}015$ mmol l^{-1} from 0 to $1{\cdot}5$ mmol l^{-1};
$0{\cdot}03$ mmol l^{-1} from $1{\cdot}5$ to $3{\cdot}0$ mmol l^{-1}

LEVEL II

4.6.2. Volumetric; with indicator

Principle

To the chloride is added an excess of $AgNO_3$ in *acid* solution, in order to prevent the co-precipitation of the Ag-salts of CO_3^{2-}, PO_4^{3-}, SiO_3^{2-} etc. (Such precipitation is a disadvantage of Mohr's method). Without removing the AgCl, the excess Ag^+ is titrated with CNS^- in the presence of Fe^{3+} and nitrobenzene. (Volhard's method).

Reagents

A **Standard NaCl, 0·0100 M** See § 4.6.1

B **$AgNO_3$, 0·1 M = 100 mmol l^{-1}**
Dissolve 8·5 g of $AgNO_3$ (dry, A.R.) in H_2O and dilute to 500 ml. Standardize against NaCl as described in the procedure below.

C **HNO_3, 1 + 1**
Mix equal volumes of HNO_3 (s.g. 1·42) and H_2O. Boil until colourless and store in a glass stoppered brown bottle.

D **$Fe_2(SO_4)_3$, 25%**
Dissolve 25 g of $NH_4Fe(SO_4)_2.12H_2O$ in about 100 ml of H_2O to which 5 drops (0·25 ml) of HNO_3 (C) have been added.

E **KCNS, 0·02 M = 20 mmol l^{-1}**
Dissolve 2 g of KCNS in 1 litre of H_2O. Standardize against $AgNO_3$ as described in the procedure below.

F **Nitrobenzene, pure (poisonous)**

Procedure

Mix 100 ml of the sample containing not more than 35 mg Cl^- (1 mmol Cl^-), or an aliquot diluted to 100 ml, with 5 ml HNO_3 (C). Add $AgNO_3$ (B) from a burette, mixing vigorously, to the approximate end point. This may be recognized as the point when no precipitation occurs near the $AgNO_3$ drops. Add about 2 ml more (= excess) of $AgNO_3$, and record the volume of $AgNO_3$ added. Add 3 ml of nitrobenzene (F) and 1 ml of $Fe_2(SO_4)_3$ solution (D). Shake vigorously and titrate the excess Ag^+, after the coagulation of the precipitate, with standardized KCNS (E). At the end point a permanent reddish colour appears, which does not fade in 1 minute of intensive shaking.

For the standardization of the KCNS (E) use 100 ml of H_2O + 10·0 ml of $AgNO_3$ (B).

Interferences
I^-, Br^- and S^{2-} are measured as equivalents of Cl^-.

Calculation

Let x = concentration of Cl^- in the water sample [mmol l^{-1}]

then $$x = \frac{(C_b V_b - C_e V_e)}{V_s}$$

where C_b = concentration of reagent B [mmol l^{-1}]
V_b = volume of reagent B [ml]
C_e = concentration of reagent E [mmol l^{-1}]
V_e = volume of reagent E [ml]
V_s = volume of water sample [ml]

Let y = concentration of Cl^- in the water sample [mg l^{-1}]

then $$y = 35{\cdot}5\,x = \frac{35{\cdot}5\,(C_b V_b - C_e V_e)}{V_s}$$

Precision and accuracy
Both are limited by the accuracy of the end point detection, which is better than 0·2 ml, or 0·1 mg Cl^-. To obtain higher precision the $AgNO_3$ may be added from a pipette. This means however that the concentration of Cl^- must be known approximately before starting.

Note
The use of the commercially available standardized solutions of $AgNO_3$ and KCNS is strongly recommended.

LEVEL II

4.6.3. Potentiometric titration

Principle
The change of an Ag-electrode potential with the Ag^+ concentration in the solution, is of the same nature as the change of the potential of the glass electrode with H^+ concentration—see § 9.7 and § 9.9. The Ag-electrode can therefore be used to follow the course of the titration with Ag^+.

Reagents

A_1 $AgNO_3$, 0·1 M See § 4.6.2

A_2 $AgNO_3$, 0·02 M = 20 mmol l^{-1}
Dilute A_1 5 times.

B HNO_3, 1 + 1 See § 4.6.2

Apparatus
Silver rod indicator electrodes and reference electrodes, made of either calomel or silver-silver chloride, may be bought. (Cl^- diffuses from calomel electrodes into the sample, so the electrode must be placed in a second flask connected to the titration flask by an NH_4NO_3 or KNO_3 agar bridge. Alternatively $HgSO_4$ electrodes or even glass electrodes can be used). A normal pH meter can be used.

Procedure
Mix 100 ml of sample, containing not more than 35 mg of Cl^- (1 mmol Cl^-) if using 0·1 M $AgNO_3$, or not more than 7 mg of Cl^- (0·2 mmol Cl^-) if using 0·02 M $AgNO_3$, or an aliquot diluted to about 100 ml, with 10 drops (0·5 ml) of HNO_3 (B). Add $AgNO_3$ from a burette, at first in relatively large portions (e.g. 1/5, 2/5, 3/5 and 4/5 of the expected quantity) but approaching the equivalence point in smaller quantities, depending on the precision required. Continue beyond the equivalence point in larger steps again. Record the potential difference at each point.

Interferences
Br^-, I^- and S^{2-} are measured as equivalents of Cl^-.

Calculation
Plot potential difference E, versus volume $AgNO_3$ used, and if necessary

$$\frac{\Delta E}{\Delta V} \quad \text{and} \quad \frac{\Delta^2 E}{\Delta V^2}$$

versus volume of $AgNO_3$ used. The last curve gives the greatest precision.
Then follow the calculation of § 4.6.4.

Precision and accuracy
These depend on the apparatus used and the quantity of Cl^- present. The accuracy can be 0·2% if a microburette is used.

Sensitivity
The sensitivity is about 1–2 mg l^{-1} of Cl^-. In this range and below use the method of § 4.6.4.

4.6.4. Conductometric titration for low Cl^- concentrations

Principle
See § 9.1 and fig. 9.3B.

Reagents
As for § 4.6.3.

Apparatus
A normal conductance meter can be used. Meters with a titration attachment are preferable for accurate work.

Procedure
As for § 4.6.3, but also record the conductance well beyond the end point. Note that if the curve before the equivalence point is nearly horizontal, then the precision

of the conductometric titration depends mostly on the points *beyond* the equivalence point. Points around the end point are of no use.

Interferences
As for § 4.3.3.

Calculation
Plot the conductance versus volume of $AgNO_3$. The equivalence point is given by the point of intersection of the two lines

Let x = concentration of Cl^- in the water sample [mmol l^{-1}]

then $$x = \frac{C_a V_a}{V_s}$$

where C_a = concentration of reagent A_2 [mmol l^{-1}]
V_a = volume of reagent A_2 [ml]
V_s = volume of sample [ml]

Let y = concentration of chloride in the water sample [mg l^{-1}]

then $$y = 35{\cdot}5\,x = \frac{35{\cdot}5\,C_a V_a}{V_s}$$

Precision and accuracy
As for § 4.6.3.

LEVEL III

4.6.5. Potentiometric

The same as in level II, but with an automatic recorder and motorburette or an automatic titrator. The procedure § 4.6.3 can easily be automated.

4.7 SULPHATE

No satisfactory direct colorimetric method for measuring sulphate concentration exists. The gravimetric method can be made very accurate, but is then time consuming. Most routine methods are based on the precipitation of sulphate after adding Ba^{2+}. The estimate is then made by measuring the turbidity or by measuring the excess of Ba^{2+}.

LEVEL I

4.7.1. Turbidimetric

Principle
SO_4^{2-} is precipitated with Ba^{2+} in an acid solution. It is assumed that it is possible to obtain $BaSO_4$ crystals of uniform size. (Glycerol–ethanol solution is added as a stabilizer.) Absorbance is measured with a (battery operated) colorimeter or

As for S 4.3.4.

then, y = [illegible]

As for [illegible]

turbidimeter. Any wavelength between 380 and 420 nm, or white light, can be used.

Reagents

A Standard $\frac{1}{2}$ H_2SO_4, 0·0200 M (= 20·0 mmol l^{-1} = 961 mg l^{-1} SO_4^{2-})
Dilute 5 ml 4 M $\frac{1}{2}$ H_2SO_4 (§ 1.3.4) to 1 litre.
Standardise with 0·100 M NaOH (§ 1.3.4).

B NaCl–HCl solution
Dissolve 240 g of NaCl in 900 ml of H_2O. Add 20 ml of HCl (A.R., s.g. 1·19) and dilute to 1 litre.

C $BaCl_2.2H_2O$ (dry, 20–30 mesh crystals)

D Glycerol–alcohol solution
Mix 1 volume of glycerol with two volumes of ethanol.

Procedure
Mix 50·0 ml of filtered sample containing not more than 10 mg l^{-1} SO_4^{2-} (about 0·2 mmol l^{-1} $\frac{1}{2}$ SO_4^{2-}), or an aliquot diluted to 50·0 ml, with 10·0 ml each of solution (B) and (D). Measure the absorbance against a H_2O blank in a colorimeter at any wavelength between 380 nm and 420 nm or in a turbidimeter with white light.

Add 0·15 g of $BaCl_2$ (a 'constant spoonful') and place on a magnetic stirrer for 30 minutes, or shake constantly and in a repeatable manner for 5 minutes exactly. Measure the absorbance after 30 minutes. The absorbance due to SO_4^{2-} is obtained by difference.

Calibration and calculation
Prepare a calibration curve in the range of 1 to 10 mg l^{-1} SO_4^{2-} (about 0·02 to 0·2 mmol l^{-1} SO_4^{2-}) using dilutions of H_2SO_4 (A). It is desirable to make this calibration curve with the lake water itself (standard addition).

1 mmol $\frac{1}{2}$ SO_4^{2-} = 48·05 mg SO_4^{2-} = 16·0 mg SO_4-S.

Interferences and errors
In highly acidified fresh water, SO_4^{2-} is the only ion which forms a precipitate with Ba^{2+}. Coloured and suspended compounds interfere, while organic compounds may alter the crystal size.

Precision and accuracy
The precision will be about 10%, but under favourable conditions (e.g. dilution of the sample) it may be better. The use of monochromatic light and a magnetic stirrer improves the precision. Standard addition will improve the accuracy.

4.7.2. By calculation (from the conductivity)
The sulphate concentration can sometimes be calculated from the conductivity of the sample and the concentrations of NO_3^- and Cl^-. See Mackereth (1955).

LEVEL II

4.7.3. Complexometric (Tentative)

Principle

An accurately measured (excess) amount of Ba^{2+} is added to the sample. $BaSO_4$ precipitates, and the remaining Ba^{2+} is back-titrated with Na_2EDTA; see § 4.3. Interference may occur from Ca^{2+} or Mg^{2+}. If SO_4^{2-}, and Ca^{2+} plus Mg^{2+} are present in high concentration, the cations are best removed by sorption on a cation exchange resin. The effluent SO_4^{2-} may then be determined. If the concentration of Ca^{2+} plus Mg^{2+} is high relative to that of SO_4^{2-}, then it may be better to separate the cations and anions with an anion exchange resin. These retain the SO_4^{2-}, which is subsequently eluted and estimated. An additional advantage of this method is that the SO_4^{2-} may be concentrated from a large sample volume.

Organic compounds, which may also interfere, are partly destroyed by boiling the eluate (from the anion exchange resin) with $HClO_4$.

The anion exchanger method is given here in detail; the cation exchanger method only in outline.

Reagents

A_1 ½ Na_2EDTA, 0·100 M = 100 mmol l^{-1}

A_2 ½ Na_2EDTA, 0·040 M = 40 mmol l^{-1}

Dilute 0·1 M Na_2EDTA (See § 1.3.4).
(A_1 50·0 ml → 100·0 ml; A_2 50·0 ml → 250 ml)
See note on H_2O § 1.3.2.

B Mg–Na_2EDTA solution, 0·250 M = 250 mmol l^{-1}

Dissolve 98·5 g of Mg–Na_2EDTA in 1000 ml of H_2O.

C HCl, 1 M

Dilute 4 M HCl (see § 1.3.4) (0·25 litre → 1 litre).

D $HClO_4$, 60% (warning: potentially dangerous, see Everett, 1972)

E NaOH, 1 M

Dilute 10 M NaOH (see § 1.3.4).

F NH_4OH, 2·5 M

Dilute 25% NH_4OH (0·2 → 1 litre).

G_1 ½ $BaCl_2$, 0·02 M = 20 mmol l^{-1}
G_2 ½ $BaCl_2$, 0·10 M = 100 mmol l^{-1}

Dissolve 2·5 g (G_1) or 12·5 g (G_2) of $BaCl_2.2H_2O$ in 1 litre of H_2O.

Standardize against the EDTA solution as described in the procedure below.

H_1 Buffer solution
See Ca^{2+} plus Mg^{2+} determination, § 4.3.1, solution C_1.

H_2 Buffer solution
See Ca^{2+} plus Mg^{2+} determination, § 4.3.1, solution C_2.

I Eriochrome Black T indicator (solid)
See § 4.3.1 (G).

J KCN solution, 0·1% (highly poisonous)
Dissolve 0·1 g of KCN in 100 ml of H_2O. Keep in a brown glass bottle. Dispense from a dropping bottle.

K Methyl orange, 0·05% See § 3.4.1

L Amberlite I.R.A. 68 anion exchange resin column in HCO_3^- form
Details of resin column packing and management are given in most books dealing with ion exchange. The most important point is to prevent air getting into the prepared column. This may be done by taking a capillary tube from the outlet of the resin bed to just above the top of the resin bed (as in fig. 5.2).

Prepare a column, about 1–2 cm diameter with a resin bed about 20 cm deep. To do this, close the outlet and pour in a slurry of resin and H_2O. The resin will settle to form a bed without air bubbles.

The resin must be converted to the HCO_3^- form by passing through it about 100 ml of H_2O saturated with CO_2 (bubble CO_2 for about half an hour). The bed may be used repeatedly, but should be regenerated to the HCO_3^- form before each determination.

Procedure 1
Titrate a 50 ml filtered subsample with HCl (C) to the methyl orange (K) end point. To a second filtered subsample of the same volume add the same amount of HCl (C), and then an accurately measured (excess) amount of $BaCl_2$ (G_1).

Boil for 1 minute and cool to room temperature. Titrate the excess of Ba^{2+} (plus Ca^{2+} and Mg^{2+}) as described for the Ca^{2+} plus Mg^{2+} determination (§ 4.3.1 or § 4.3.3).

Interference and errors
In the acidified fresh water sample SO_4^{2-} ions alone precipitate with Ba^{2+}. The concentrations of Ca^{2+} and Mg^{2+} can, however, be so high that the estimation becomes unreliable. Humic substances or iron or both also affect the result. Ion-exchangers can be used to overcome interferences from these causes. Ca^{2+} and Mg^{2+} can be removed by passing 100 ml of the sample over a strong cation-exchanger in H^+ form (washed free of acid with H_2O), and discarding the first 25 ml of eluate. The following 50 ml may be used as the acidified sample in the procedure of § 4.7.3.

Calculation
Let x = concentration of $\frac{1}{2}$ SO_4^{2-} in the water sample [mmol l^{-1}]

Then $$x = C_d + \frac{(C_b V_b - C_e V_e)}{V_s}$$

where C_d = concentration of $\frac{1}{2}$ Ca^{2+} + $\frac{1}{2}$ Mg^{2+} [mmol l^{-1}] estimated for exampl by procedure § 4.3. If ion exchangers are used then $C_d = 0$
C_b = concentration of $\frac{1}{2}$ $BaCl_2$ [mmol l^{-1}]
V_b = volume of $\frac{1}{2}$ $BaCl_2$ added [ml]
C_e = concentration of $\frac{1}{2}$ Na_2EDTA [mmol l^{-1}]
V_e = volume of $\frac{1}{2}$ Na_2EDTA [ml]
V_s = volume of water sample [ml]

Let y = concentration of SO_4^{2-} in the water sample [mg l^{-1}]

Then $$y = 48{\cdot}1\,x = 48{\cdot}1\left\{C_d + \frac{(C_b\,V_b - C_e\,V_e)}{V_s}\right\}$$

Ca^{2+} plus Mg^{2+} may be removed by procedure 2, which increases the SO_4^{2-} concentration if samples larger than 100 ml are used.

Procedure 2

Pass a suitable sized sample (usually between 100 to 500 ml) over the anion exchanger in the HCO_3^- form. The flow rate should be about 1 ml per 10–20 sec Elute with 100 ml of 2·5 M NH_4OH (F). Boil in 250 ml covered beakers with 2–5 drops (0·1–0·2 ml) of 1 M NaOH (E) till the NH_3 has been removed. Use anti bumping glass rods. Acidify with $HClO_4$ (D) to pH = 1·5–2·0. Heat to boiling and add slowly exactly 4·0 ml of $BaCl_2$ (G_2) from a piston microburette. Boil for 10 min. Cool and adjust the pH to 9 with 2·5 M NH_4OH (F). Add from a dropping bottle 5 drops of KCN (J), 4 ml of Mg EDTA (B) and buffer solution (H_2) to pH = 10. Add about 50 mg of the indicator (I) and titrate at once with 0·100 M $\frac{1}{2}$ Na_2EDTA (A_1). Just before the end point, add more indicator if need be. Titrate from red till there is no further change in the blue colour. Make a blank using H_2O instead of the sample.

Calculation

Let x = concentration of $\frac{1}{2}$ SO_4^{2-} in the water sample [mmol l^{-1}]

then $$x = \frac{C_e\,(V_{eb} - V_{es})}{V_s}$$

where C_e concentration of $\frac{1}{2}$ Na_2EDTA [mmol l^{-1}]
V_{eb} = volume of $\frac{1}{2}$ Na_2EDTA used in the blank titration [ml]
V_{es} = volume of $\frac{1}{2}$ Na_2EDTA used in the sample titration [ml]
V_s = volume of water sample

Let y = concentration of $\frac{1}{2}$ SO_4^{2-} in the water sample [mg l^{-1}]

Then $$y = 48{\cdot}1\,x = \frac{48{\cdot}1\,C_e\,(V_{eb} - V_{es})}{V_s}$$

Note

If sufficient SO_4^{2-} is present procedure § 4.7.4 may be tried after acidification with $HClO_4$. In this case, use as little NaOH as possible.

4.7.4. Potentiometric (Tentative)

(Only for SO_4^{2-} exceeding 50 mg l^{-1}).

Principle

Pb^{2+} forms a precipitate with SO_4^{2-}. Ethanol is added to reduce the solubility of the $PbSO_4$. When all the SO_4^{2-} ions are precipitated Pb^{2+} will precipitate ferrocyanide (added to the sample) as lead-ferrocyanide, causing a sharp change in potential of an indicator electrode. Pt and Ag-electrodes are used.

Reagents

A $\frac{1}{2}$ H_2SO_4, 0·0200 M See § 4.7.1.

B $\frac{1}{2}$ $Pb(NO_3)_2$, 0·025 M = 25 mmol l^{-1}
Dissolve 4·14 g of $Pb(NO_3)_2$ in 1 litre of H_2O.

Standardize against 0·0200 M $\frac{1}{2}$ H_2SO_4 (Solution A) as described in § 4.7.4, procedure.

C $K_4Fe(CN)_6$ solution, 0·005 M
Dissolve 0·21 g of $K_4Fe(CN)_6.3H_2O$ (A.R.) in about 100 ml of H_2O.

D $K_3Fe(CN)_6$ solution, 0·1 M
Dissolve 33 g of $K_3Fe(CN)_6$ (A.R.) in 1 litre of H_2O.

Procedure

Put 25·0 ml of slightly acidified sample into a 100 ml titration flask fitted with a Pt-electrode (0·5 mm diameter) as reference electrode and a Ag-electrode (3–4 mm rod). Add an equal volume of 96% ethanol, 0·1 ml of solution (C) and 1·0 ml of solution (D). Stir with a magnetic stirrer, and add from a 5 ml micro-burette the $Pb(NO_3)_2$ (B) in portions of 1 ml at first, and then in portions of 0·1 ml when the end point is near.

Interferences

Phosphate and Ca^{2+} may interfere if present in high concentrations.

Precision and accuracy

Unknown.

Calculations

Plot $\frac{\Delta E}{\Delta V}$ or $\frac{\Delta^2 E}{\Delta V^2}$ against volume as described in § 9.12.

Let x = concentration of $\frac{1}{2}$ SO_4^{2-} in the water sample [mmol l^{-1}]

Then $$x = \frac{C_b V_b}{V_s}$$

where C_b = concentration of reagent B [mmol l^{-1}]
V_b = volume of reagent B [ml]
V_s = volume of water sample [ml]

Let y = concentration of $\frac{1}{2}$ SO_4^{2-} in the water sample [mg l^{-1}]

Then $$y = 48{\cdot}1\,x = \frac{48{\cdot}1\,C_b V_b}{V_s}$$

LEVEL III

4.7.5. Sulphate by AAFS

Principle
As for Level II § 4.7.3. An accurately measured excess of $BaCl_2$ is added. $BaSO_4$ precipitates and is allowed to settle overnight. Then the concentration of Ba^{2+} in the supernatant is estimated by AAFS.

Ionisation of Ba^{2+} interferes but may be suppressed by adding NaCl (3 g l^{-1} of Na^+).

Wavelength 553·6 nm. Reducing nitrous oxide-acetylene flame. The detection limit for Ba^{2+} is about 0·1 mg l^{-1}.

4.8 TOTAL AND DISSOLVED SULPHIDE

LEVEL I

4.8.1. Titrimetric

Principle
The S^{2-} is precipitated with $CdCl_2$ in the same type of bottle as used for the Winkler method of O_2 determination (see § 8.1). When the precipitate has settled the supernatant is removed and the CdS is dissolved in an acid iodine solution. The excess iodine is titrated with $S_2O_3^{2-}$

$$H_2S + I_2 \rightarrow S + 2H^+ + 2I^-$$

Reagents

A HCl, 4 M See § 1.3.4

B $Na_2S_2O_3$ solution, 0·025 M = 25 mmol l^{-1} See § 8.1.2

C $CdCl_2$, 2% See § 4.8.2

D Iodine solution, $\frac{1}{2}$ I_2 0·025 M = 25 mmol l^{-1}
Dissolve 20 g of KI in 50 ml of H_2O and add 3·17 g of I_2. After the iodine has dissolved, dilute to 1 litre and standardize against standardized $Na_2S_2O_3$ with starch as indicator (see § 8.1.2).

Procedure
Completely fill a glass stoppered bottle with the water sample in the field, following the procedure described in § 8.1.2. The volume of the bottle should be about 110 ml, and should be known. Add 1 ml $CdCl_2$ solution (C). Take the sample to the laboratory and allow to stand for 24 or 48 hours. Decant the supernatant (for example with a pipette connected to a vacuum pump). Dissolve the precipitate in an exactly known small volume of iodine solution (D) and 5 ml HCl (A). Titrate the excess iodine with standardized $S_2O_3^{2-}$ as described in procedure § 8.1.2.

Interferences
If much organic matter is present the precipitate should be collected on a membrane filter and washed with H_2O.

Calculation

Let x = concentration of $\frac{1}{2}$ S^{2-} in the water sample [mmol l^{-1}]

then $$x = \frac{(C_i V_i - C_t V_t)}{V_s - 1}$$

where C_i = concentration of $\frac{1}{2}$ I_2 [mmol l^{-1}]
V_i = volume of iodine solution used [ml]
C_t = concentration of $\frac{1}{2}$ $Na_2S_2O_3$ [mmol l^{-1}]
V_t = volume of $\frac{1}{2}$ $Na_2S_2O_3$ [ml]
V_s = volume of water sample

Precision and accuracy

The precision and the accuracy are estimated at 5–10% and depend on the care with which the different steps are carried out.

LEVEL II

4.8.2. Colorimetric

Principle

H_2S and N,N-dimethyl-p-phenylenediamine (I) (= p-amino dimethylaniline) are converted to methylene blue (II) in the presence of $FeCl_3$ and in suitable physical conditions.

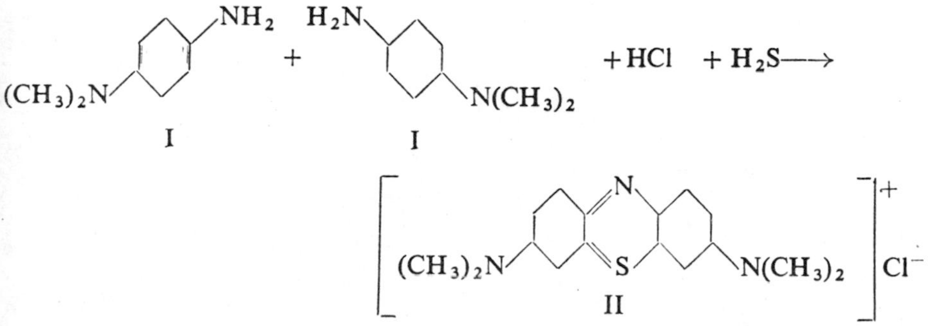

I I II

A distinction between dissolved and total H_2S can be made by using a filtered and an unfiltered sample.

Storage

If necessary the S^{2-} in a sample can be stored as a precipitate by adding 2 ml of solutions (C) or (F) per litre of sample (for total sulphide) or of filtrate (for dissolved sulphide). Each precipitate must then be transferred quantitatively into a 100 ml volumetric flask.

Reagents

A_1 Standard Na_2S solution

Rinse quickly with H_2O about 0·6 g of large crystals of $Na_2S.9H_2O$. Discard the washings. Dissolve the washed crystals in about 500 ml of recently boiled H_2O which has cooled out of contact with the air. Fill the container completely (to exclude gas bubbles), stopper, and mix.

Mix 200 ml of H_2O with 50·0 ml 0·025 M iodine solution (§ 4.8.1), 25 ml of 0·1 M HCl (see § 1.3.4) and 20·0 ml of solution (A_1). Titrate the excess iodine with thiosulphate (see § 8.1.2). 2·5 ml of iodine solution is equivalent to 1·00 mg of S^{2-} (0·0625 mmol $\frac{1}{2}$ S^{2-}).

A_2 Diluted Na_2S standard
Dilute an aliquot of solution (A_1) containing 10·0 mg of S^{2-} (about 0·6 mmol $\frac{1}{2}$ S^{2-}) to 500 ml with O_2-free H_2O. Stopper at once and mix. Avoid contact with air.

Both solutions are extremely unstable and must be used immediately.

B Stock solution of N,N-dimethyl-p-phenylenediamine
Add 50 ml of H_2SO_4 (A.R.) (s.g. 1·84) to 30 ml of H_2O (*Danger*, see § 1.3.3), and cool. Dissolve in this solution 20 g of the redistilled amine or 27 g of the amine sulphate. Stir and make up to about 100 ml with H_2O.

B_1 Test solution, 0·5%
Mix 25 ml of the stock solution (B) with 975 ml H_2SO_4 (1 + 1) (see § 1.3.4).

B_2 Test solution, 0·2%
Mix 10 ml of the stock solution (B) with 990 ml H_2SO_4 (1 + 1) (see § 1.3.4).

Do not use solutions B_1 and B_2 which have been mixed for more than 24 hours.

C $CdCl_2$ solution, 2%
Dissolve 20 g of $CdCl_2$ in 1 litre of H_2O.

D_1 $FeCl_3$ solution, 0·1 M
Dissolve 27 g of $FeCl_3.6H_2O$ in 1000 ml of HCl (1 + 1).

D_2 $FeCl_3$ solution, 0·02 M
Dilute before use one volume of solution D_1 with 4 volumes of H_2O.

E $(NH_4)_2HPO_4$ solution, 3 M
Dissolve 400 g $(NH_4)_2HPO_4$ in 800 ml H_2O.

F Zn acetate solution, 1 M
Dissolve 240 g $Zn(CH_3COO)_2.2H_2O$ in 1 litre of H_2O.

Procedure
Put 90·0 ml of sample or the CdS or ZnS precipitates into a 100 ml volumetric flask, and add 5 ml (B_1) or (B_2) (neither more than 24 hours old). Use (B_1) if 0·2 to 20 mg l^{-1} S^{2-} and (B_2) if 0·05 to 5 mg l^{-1} S^{2-} is expected. Mix well and add 1 ml of freshly diluted $FeCl_3$ solution (D_2) with gentle shaking. Dilute to 100 ml. Measure the absorbance of reagent blank, standards and samples in suitable cells at a wavelength as close as possible to 745 nm after not less than 15 minutes but not more than 2 hours.

If a separation between dissolved and particulate S^{2-} is required the filtration must be carried out with exclusion of O_2, for example under N_2. If vacuum filtration is to be used the sample must be made strongly alkaline beforehand, to prevent loss of H_2S.

The yellow colour of the iron may be destroyed by adding a few ml of solution (E).

Calibration and calculation
Make a calibration curve covering the range 0·05 to 20 mg S^{2-}; 0·25 to 1·25 mmol $\frac{1}{2}$ S^{2-}.

Interferences
Organic sulphides do not interfere.

Strong reducing agents prevent the formation of the colour. High S^{2-} concentrations (above several hundred mg l^{-1}) may inhibit the reaction. Dilute in this case. SO_3^{2-} does not interfere up to 10 mg l^{-1}.

Aeration of samples and standard solutions leads to oxidation and volitalization and must therefore be avoided.

Precision and accuracy
The standard deviation is unknown. The accuracy is about 10% and depends mainly on the accuracy of the standard curve.

The method is sensitive to 0·005 mg l^{-1} S^{2-}.

4.9 CARBONATE AND BICARBONATE

See § 3.4 and § 3.5.

4.10 CALCULATION OF IONIC BALANCE AND CONDUCTIVITY

LEVEL II

4.10.1. Ionic balance

The sum of concentrations of positive charges must equal the sum of concentrations of negative charges. If it is supposed that all major constituents have been estimated the calculation of ionic balance therefore provides a useful check of overall accuracy. Comparison of observed with calculated conductivity (§ 4.10.2) provides another check. Used together these checks act like diagonal bracing structures, making a single firm reliable unit from an uncertain framework.

For the ionic balance calculation the total alkalinity (§ 3.4) may be used, but for conductivity calculations HCO_3^-, $\frac{1}{2}$ CO_3^{2-} and perhaps organic acid concentrations should be known separately. The concentration of H^+ or OH^- may be estimated with sufficient accuracy from the pH. Between pH 4·5 and 9·5 both are negligible. For very weak solutions the conductivity of the water itself should be subtracted.

An example of the calculation is shown in § 4.10.3. In an ionic balance there is usually a difference of 2% to 5% as a result of analytical errors, even when all the major constituents have been analysed. A smaller imbalance may result from compensating errors; a larger one suggests the presence of significant amounts of an unanalysed constituent. For example NH_4^+ may be of importance in anaerobic waters. (The anions of weak acids, including most organic acids, are included in the alkalinity estimate).

4.10.2. Calculated conductivity

The calculated conductivity is approximately the sum for all charged constituents of the concentration multiplied by the molar conductivity particular to that

constituent (Table 4.1). If the measured conductivity is not much above 100 μS cm^{-1} this calculation should be inaccurate by no more than 5% in most cases. The sample can be diluted by a known amount to meet this condition. The calculation is less reliable if there are relatively high concentrations of 'silicate' or 'iron' or 'sulphide' because the state of these constituents depends markedly on the pH and state of oxygenation of the sample.

Table 4.1. Molar conductivity at infinite dilution [μS cm^{-2} mol^{-1}]

	18°C	25°C		18°C	25°C
H^+	314	350	OH^-	172	198
Na^+	44	50	Cl^-	66	76
K^+	65	74	NO_3^-	62	71
NH_4^+	65	74	HCO_3^-		45
$\frac{1}{2}$ Ca^{2+}	51	60	$\frac{1}{2}$ CO_3^{2-}		69
$\frac{1}{2}$ Mg^{2+}		53	$\frac{1}{2}$ SO_4^{2-}	68	80
$\frac{1}{2}$ Fe^{2+}		54	$SiO(OH)_3^-$		68
$\frac{1}{3}$ Fe^{3+}		52	small organic acids	about 32–40	
$\frac{1}{3}$ Al^{3+}		78			

4.10.3. Ionic balance and conductivity example

This example is an analysis of the surface water of an African lake (Denny, 1972).

Cations Constituent	Concentration mM	Contribution to Conductivity μS cm^{-1}	Anions Constituent	Concentration mM	Contribution to Conductivity μS cm^{-1}
Na^+	0·66	33	Cl^-	0·82	63
K^+	0·15	11	$\frac{1}{2}$ SO_4^{2-}	0·14	11
$\frac{1}{2}$ Ca^{2+}	0·87	52	*HCO_3	1·95	87
$\frac{1}{2}$ Mg^{2+}	0·85	45			
‡'NH_4^+'	0·09	7	‡'$SiO(OH)_3^-$'	0·02	1
†H^+	<0·001		†OH^-	<0·001	
pH = 8					
	2·62			2·93	

*Total alkalinity which at pH 8 is virtually all HCO_3^-, see Fig. 9.1.
†Calculated from pH.
‡Calculated as NH_4^+ and '$SiO(OH)_3^-$' which at this pH overestimates both.

The calculated conductivity at 25°C is 310 μS cm^{-1}; that measured was 258 μS cm^{-1}.

The ionic balance and conductivity check might be considered just adequate at Level I, which was all that the analyst needed for his purpose. At Level II one would expect agreement to within 2% in a satisfactory analysis.

CHAPTER 5

MINOR CONSTITUENTS*

NITROGEN, PHOSPHORUS AND SILICON COMPOUNDS

NITROGEN COMPOUNDS

5.1 AMMONIA, NITRATE AND NITRITE

Note
Special attention should be given to ensure that no NH_3 from the environment is taken up during these analyses, particularly by the H_2SO_4. For example, smoking should be prevented in the laboratory. Ammonia bottles should not be kept or opened in the laboratory. Glassware should be acid-washed, rinsed well with ammonia-free water, and stored stoppered.

LEVEL II

Colorimetric

5.1.1. Ammonia

5.1.2. Nitrate and nitrite

Principle

NH_4^+ reacts with K_2HgI_4 in strongly alkaline conditions to form a brown coloured substance. In nearly all fresh waters interfering substances are present which make a steam distillation of the NH_3 obligatory. This distillation can also be used to concentrate the NH_3.

NO_3^- (and NO_2^-) are completely reduced to NH_3 by Devarda's alloy under strongly alkaline conditions and may then be determined as NH_3. The NO_3^- and NO_2^- determination may be made later on the same sample as was used for NH_3 determination.

Apparatus
There are a number of satisfactory designs of apparatus for distilling off NH_3. That described here is a modified Parnas–Wagner still (see Fig. 5.1). Simpler apparatus can also be used, but will not usually allow the determination of NH_3, the reduction of NO_3^- and the determination of organic nitrogen in one sample.

Reagents

H_2O
Ammonia-free water should be used for the preparation of reagents and blanks, and for the dilution of solutions. Traces of ammonia can be removed by distillation

* MINOR refers to concentration, not to biological importance.

of the water after the addition of a small quantity of H_2SO_4, or by passing the water through a cation exchange resin in the Hydrogen form (e.g. Amberlite IR 120 H or Permutit Zeocarb 225). Care should be taken if a resin is used, because some resins which produce water with a very low conductivity may release organic substances which interfere with the NH_3 determination.

A_1 $(NH_4)_2SO_4$ standard solution, 200 μg ml^{-1} of NH_3–N
Dissolve 0·9433 g of $(NH_4)_2SO_4$ (A.R., dried in a desiccator) in 1000 ml of H_2O. This solution contains 200 μg ml^{-1} of NH_3–N. Standardize with procedure § 5.2.2.

A_2 $(NH_4)_2SO_4$ standard solution, 2 μg ml^{-1} of NH_3–N
Dilute solution A_1 100 times. Solution A_2 contains 2 μg ml^{-1} of NH_3–N.
Prepare fresh as required.

B KNO_3 standard solution, 2 μg ml^{-1} of NO_3–N
Dissolve 0·3611 g of KNO_3 (A.R. dried at 105°C) in 250 ml of H_2O. Dilute an aliquot 100 times (10·0 ml → 1000 ml). This diluted solution contains 2 μg ml^{-1} of NO_3–N.

C H_2SO_4, 0·02 M
Add 2 ml of H_2SO_4 (1 + 1) (A.R.) (see § 1.3.4) to 1 litre of H_2O.

D NaOH, 5 M
Dilute 10 M NaOH (see § 1.3.4).

E $Na_2B_4O_7.10H_2O$ (saturated solution)
Add 4 g of $Na_2B_4O_7.10H_2O$ (A.R.) to 100 ml of H_2O. Heat until all crystals are dissolved and store at room temperature. A small amount of precipitate will appear.

F Nessler's reagent

F_1
Dissolve 25 g of HgI_2 (red) and 20 g of KI in 500 ml of H_2O.

F_2
Dissolve 100 g of NaOH in 500 ml of H_2O. These reagents can be stored for several months in air-tight bottles. Glass stoppered bottles are not suitable.

Mix 1 vol. F_1 + 1 vol. F_2, and keep in a refrigerator.

G Devarda's alloy (A.R.)
Commercially available. If blanks are high due to the Devarda's alloy it must be purified by heating for one hour at 120–150°C in an open dish.

Procedure

5.1.1. Ammonia
Put a 50 ml volumetric flask below the condenser outlet of the distillation apparatus if 50–500 μg of NH_3–N is expected. Use a 25 ml volumetric flask if only 5–100 μg of NH_3–N is expected.

Close tap C. Put the sample in the distillation apparatus through the funnel and tap A. Add 1 ml of buffer solution (E). Close tap A immediately. The outlet of the condenser dips just into 2·5 ml of 0·02 M H_2SO_4 (C) in the volumetric flask. Close tap B and pass steam through the apparatus until 40 ml (or 20 ml if the smaller flask is used) of distillate has been collected. Lower the volumetric flask and collect a further 2 to 5 ml of distillate. If NO_3^- is to be determined subsequently (§ 5.1.2), leave the sample in the still by opening tap B and then shutting off the steam supply. If the sample is no longer required however it may be removed by simply shutting off the steam supply. The sample will suck back into the outer container and may be removed by opening tap B. Make the distillate up to volume and mix. Take 20 ml of the distillate (or an aliquot containing not more than 100 μg of NH_3–N and diluted to 20 ml) in a tube 20 cm × 3 cm and add 1·00 ml of Nessler's reagent (F) very slowly, while swirling. Cover with, for example, a watch glass.

Run a blank (using twice distilled H_2O) before and after each series of samples, and standards covering the range 10 to 100 μg of NH_3–N. The same curve should fit standards both before and after distillation. Measure the absorbance of the reagent blanks, standards and samples, against water, in suitable cells using any wavelength between 380 and 430 nm. Reagent blanks in one series should not differ more than 5%. Daily variations may be two or three times as much.

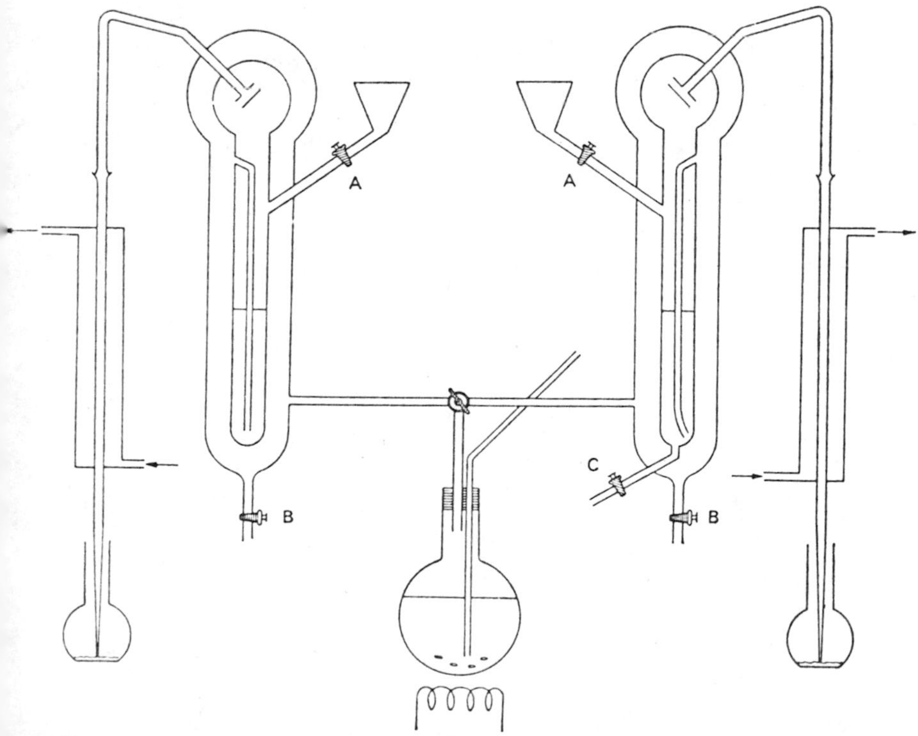

Figure 5.1. Modified Parnas–Wagner distillation apparatus. Left side for NH_3 only. Right side for sample recovery after Devarda reduction or NH_3 determination.

5.1.2. Nitrate (and nitrite)

Add about 0·2 g of Devarda's alloy (G) to the sample—after NH_3 has been removed by distillation—and then add slowly 5 ml of 5 M NaOH. Put a volumetric flask containing H_2SO_4 (C) below the condenser outlet (as described for NH_3–N determination). Close tap A. Leave tap B open and pass steam through the outer container (but not through the sample) for 10 minutes. The sample is thus heated and H_2 gas is generated. Because the condenser outlet is narrow, pressure inside the apparatus increases, and sample may be forced over into the outer steam jacket. To prevent this happening the tap B may have to be closed partly, but not so much as to force steam through the sample. During this procedure the NO_2^- and NO_3^- are reduced to NH_3, which is later distilled off by completely closing tap B.

Determine the NH_3 as described above.

Calibrate against the NH_3 standard curve, and use KNO_3 to check that reduction is stoicheiometric. The sample can be recovered through tap C for determination of organic nitrogen compounds (§ 5.5.1).

Calculation

The amount of the NO_2^- is determined according to procedure § 5.4 and must be subtracted to give the amount of NO_3^-.

Note

Blanks and samples containing Nessler's reagent and the stock of the mixed reagent itself must be exposed to the air as little as possible. With the mixed reagent, blanks are higher but the colour of the samples is more stable than with reagent mixed just before use.

Interferences

In polluted water (e.g. water polluted by sewage, water from fish ponds etc.) nitrogen compounds may occur which hydrolyse during the distillation procedure (§ 5.1.1 or § 5.1.2).

If these compounds are present they will hydrolyse in part during the first distillation step (2), but some may hydrolyse slowly, and NH_3 will be found in a subsequent distillate (3). Consider this NH_3 (3) to be organically bound and estimate the inorganic NH_3 in the first distillate by subtracting the NH_3 in distillate (3) from that in distillate (2).

Alternatively, try the procedure § 5.2.4.

When compounds which hydrolyse during the reduction step are present, add to the main procedure a distillation step using 5 M NaOH, before the reduction step takes place. Collect this distillate. Then add Devarda's alloy to the sample (carefully), and measure the NH_3 in both distillates.

The following scheme should deal with these situations.

Determination of NH_3–N

(1) Add 1 ml of borax to the sample and
(2) collect distillate; analyse for NH_3 and
(3) collect in the meanwhile another distillate and analyse for NH_3
(4) If NH_3 is present hydrolysable products are present, GO BACK TO step 3 (continue distillation).
If no NH_3 is present GO ON TO step 5

Determination of NO_3–N+NO_2–N

(5) Add 10 ml of NaOH to the sample in the still
(6) Collect distillate, analyse for NH_3

(7) If NH_3 is present, decomposable products are present, GO BACK TO step 6 (continue distillation)
If no NH_3 is present, GO ON TO step 8
(8) Add Devarda's alloy
(9) Collect distillate; analyse for NH_3

Calculations
(A) If the answer to (4) is **NO** on the first pass: Total NH_3 is given by (2) above
(B) If the answer to (4) is **YES** on the first pass: Total NH_3 is given approximately by $(2)-(3)_{1st}$
Hydrolysable NH_3 is given approximately by $2\times(3)_{1st}+(3)_{2nd}+(3)_{3rd}+(3)_{nth}$
(C) If the answer to (7) is **NO** on the first pass: $NO_3^-+NO_2^-$ is given by (9)
(D) If the answer to (7) is **YES** on the first pass: Decomposable –N is given by $(6)_{1st}+(6)_{2nd}+(6)_{3rd}+(6)_{nth}$ and $NO_3^-+NO_2^-$ is again given by (9).

After the 'reduction' distillation, and before another sample is introduced, the Devarda's alloy must be removed carefully from the still. This is laborious, and can be avoided if two apparatuses can be used; one for $NH_3+NO_2^-+NO_3^-$, and the other for NH_3 alone. This is suitable only if NH_3 and $NO_3^-+NO_2^-$ are present in approximately equal amounts.

5.1.3. Nitrate (and nitrite)

Principle
As in § 5.1.1 and § 5.1.2, but using simpler apparatus.

Apparatus
Kjeldahl flask directly connected to the top of a sloping condenser by standard glass joints, Figure 5.4. Several sets of apparatus may be held in a single rack and treated together.

Reagents
See § 5.1.1.

Procedure
Put the sample in the distillation flask and add 1 ml of buffer solution E. Distill the NH_3, collect it, and estimate it as described in § 5.1.1. When the sample has cooled, disconnect the flask and add 0·2 g Devarda's alloy and 5 ml 5 M NaOH. Quickly reconnect the flask and when the reaction has subsided, distill the newly generated NH_3, collect it, and estimate it as before. The residue in the same flask may be used for the determination of dissolved organic N § 5.5.1.

5.1.4. Nitrate

Principle
NO_3^- can be reduced to NH_3 with Zn dust in an acid medium in the cold. We give no details here: there may be circumstances in which this principle would be useful.

5.2 AMMONIA

LEVEL I

5.2.1. Method deleted from this edition

LEVEL II

5.2.2. Volumetric

Principle (modified from Milner and Zahner, 1960)
When much NH_3 is present (or is formed by the reduction of NO_3^- or by destruc tion of organic nitrogen) the determination can be carried out by titration wit NH_2SO_3H. The distillate is first collected in saturated H_3BO_3 solution. H_3BO_3 so weak an acid that it does not interfere with the acidimetric titration.

Reagents

A NH_2SO_3H, 0·0100 M = 10·0 mmol l^{-1}
Dissolve 0·971 g pure sulphamic acid in H_2O. Dilute to 1000 ml.
If preferred 0·0100 M HCl may be used in place of sulphamic acid.

B Saturated H_3BO_3 + mixed indicator
Dissolve 40 g of H_3BO_3 in 1 litre of H_2O and add 5 ml of a mixed solution c bromcresol green (0·5%) and methyl red (0·1%) in 95% ethanol. Adjust the pł with NH_2SO_3H until the bluish colour turns faint pink.

Procedure
The distillation of the NH_3 present in the sample is carried out as described i procedure § 5.1.1, but 2 ml of H_3BO_3 (B) is used in the volumetric flask instead c H_2SO_4. Titrate with 0·01 M NH_2SO_3H (A) from a microburette until the bluis colour changes back to pink. Run a blank using NH_3-free H_2O instead of th sample and titrate until the same colour as the sample.

Calculation

Let x = concentration of N in the water sample [mg l^{-1}]

Then
$$x = \frac{14{\cdot}0\, C_a (V_{as} - V_{ab})}{V_s}$$

where C_a = concentration of sulphamic acid [mmol l^{-1}]
V_{as} = volume of acid used in titration of sample [ml]
V_{ab} = volume of acid used in titration of blank [ml]
V_s = volume of water sample [ml]

Let y = concentration of N in a particulate sample [mg g^{-1}]

Then
$$y = \frac{14{\cdot}0\, C_a (V_{as} - V_{ab})}{1000\, M_s}$$

Where M_s = mass of sample [g]

Note
The titration method can only be used when sufficient NH_3–N is present to neutralize about 1 ml of 0·01 M NH_2SO_3H i.e. 0·14 mg of N per sample or 2·8 mg l^{-1} of N (using 50 ml samples). In most cases the titration cannot therefore be used for fresh water, but is suitable for samples of particulate organic nitrogen (after filtration and destruction, see § 5.5.2) and to check the NH_3–N standard solution (A_1 procedure § 5.1.1) after distillation.

:VEL II

:.3. Method deleted from this edition

:.4. Colorimetric

'nciple

ımonia reacts with phenol and alkaline hypochlorite to give indophenol blue. tassium ferrocyanide can catalyse this reaction (Liddicoat *et al.*, 1976). Sodium hloroisocyanurate is the source of hypochlorite. Artificial light is used to give timum conditions for colour development.

agents

O See § 5.1.

$(NH_4)_2SO_4$ standard solutions

§ 5.1 A_1 and A_2.

Phenol-alcohol reagent

solve 10 g A.R. phenol in 100 ml A.R. 95% v/v ethanol.

Oxidising solution

solve 0·2 g sodium dichloroisocyanurate in a solution of 1·6 g A.R. NaOH in ml H_2O. Add a solution of 20 g A.R. trisodium citrate in 40 ml H_2O. Adjust ıme to 100 ml with H_2O.

Catalyst

solve 0·5 g A.R. potassium ferrocyanide in 100 ml H_2O. Transfer to an amber :le.

Reagents B and C are stable for at least 24 hr when stored in a refrigerator but ıust be prepared daily.

aratus

lytek mercury longwave ultraviolet lamp, type MBW, with maximum energy ›ut at 365 nm (or similar u.v. source) placed in the centre of an aluminium . The all-round reflecting surface of the box allows the colour development of ral samples at the same time.

'edure

50 ml sample in a 100 ml conical flask add 2 ml phenol-alcohol reagent (B), of oxidising solution (C), and 2 ml catalyst (D), mixing between each addition. er the tops of the flasks with glass or plastic caps and place the flasks inside luminium box containing the mercury lamp. Colour development is complete) min. over a temperature range of 22–27°C. Measure the absorbance at 640 gainst distilled water using suitable cells. A calibration should be made with batch of samples. Internal standards (§ 1.2.3) should be used.

'e and precision

calibration is linear up to 20 μmol l^{-1} NH_3. Precision is about 1 to 2 %.

5.2.5. Colorimetric

Principle
Ammonia reacts with phenol and alkaline hypochlorite to give indophenol blue. Sodium nitroprusside intensifies this blue colour at room temperature. Two methods are given which utilise this principle.

Method (a) after Harwood and Kühn (1970)

Reagents (use analytical grade where possible)
H_2O See § 5.1.

A $(NH_4)_2SO_4$ standard solutions
See § 5.1 A_1 and A_2.

B NaOH, 27%
Dissolve 270 g NaOH pellets in H_2O, cool and dilute to 1 litre.

C Phosphate buffer, 5%
Dissolve 5 g of Na_3PO_4 in H_2O. Make up to 100 ml.

D Phenol stock solution
Dissolve 500 g phenol in methanol and dilute to 800 ml with methanol. Store in a refrigerator.

E Phenate reagent I
Take 15 ml of the phenol stock (D), add 0·02 g sodium nitroprusside and dilute to 100 ml.

F Alkaline bleach
Mix equal volumes (15 ml) of commercial bleach (3·5% chlorine) and 27% NaOH solution (B), and dilute to 50 ml.

Store (E) and (F) in amber bottles in a refrigerator, but allow to reach room temperature before using.

Procedure
Place a suitable aliquot of sample containing less than 50 μg N in a 25 ml volumetric flask and add 2·0 ml buffer solution (C). Dilute to about 10 ml. Add 5 ml reagent (E) with swirling, followed by 2·5 ml reagent (F), also with swirling. Dilute to the mark and mix thoroughly. Prepare blanks and standards. After about 25 minutes read the absorbance at 630 nm against the reagent blank, using a suitable cell.

Range and precision
The calibration is linear up to 2 μmol NH_3 at least. Precision is 1 to 2%.

Method (b) after Scheiner (1976)
The buffering system permits the examination of fresh and domestic wastewater, without pH correction, in the range 3–11·5.

Reagents (all reagents must be A.R.)
H_2O See § 5.1.

. $(NH_4)_2SO_4$ standard solutions

ee § 5.1 A_1 and A_2.

: NaOH, 1 M

)ilute 10 M NaOH (see § 1.3.4).

: Buffer solution, pH approximately 12·0

)issolve 30 g sodium phosphate ($Na_3PO_4.12H_2O$), 30 g sodium citrate \a$_3$C$_6$H$_5$O$_7$.2H$_2$O) and 3 g Na_2EDTA (disodium ethylenediamine tetra-acetate) H_2O and dilute to 1 litre.

Phenol-nitroprusside-buffer reagent, pH 9·1

issolve 60 g phenol completely in buffer solution (C), then add 0·2 g sodium itroprusside and make up to 1 litre with (C). Store (D) in a dark bottle, in a :frigerator and bring to room temperature before use. Prepare fresh every 3 eeks, but check after 2 weeks with at least one standard.

Alkaline hypochlorite reagent (0·08–0·11% available chlorine)

dd 30 ml commercial bleach (3·5% available chlorine) to 400 ml 1 M NaOH (B), id dilute to 1 litre. This solution is stable for at least 1 month when stored in a ark bottle in a refrigerator. Check iodometrically, as required.

pparatus

) ml volumetric flasks used exclusively for ammonia determinations. After use, ash the flasks with chromic acid (§1.2.5) or concentrated HCl and rinse oroughly. Store flasks stoppered.

ocedure

heck the pH of the sample and neutralise, if necessary. To 25 ml sample ntained in a 50 ml volumetric flask add 10 ml reagent (D). Mix by swirling. omptly add hypochlorite reagent (E), making up to the mark. Stopper the flask d mix well. After 45 min at room temperature measure absorbance at 635 nm ainst a reagent blank. Keep flasks away from bright sunlight. Prepare a calibra- n curve from standard solutions.

terferences

ardness' above 400 mg l^{-1} and NO_2^- above 15 mg l^{-1} interfere.

ecision

ecision of about 1% is claimed.

te

avitz and Gleye (1975) have established that sunlight has a marked interference the phenol hypochlorite assay. If the sample containers are wrapped with minium foil before the reagents are added and the sample is not exposed to light while being transferred to the cell, this interference is prevented.

5.3 NITRATE

LEVEL II

5.3.1. Method deleted from this edition

5.3.2. Colorimetric (tentative)

Principle

NO_3^- is quantitatively reduced to NO_2^- by a cadmium-copper couple in alkali buffered solution (pH 8). NO_2^- may then be determined as in § 5.4.1. Any NO originally present in the sample must be determined separately.

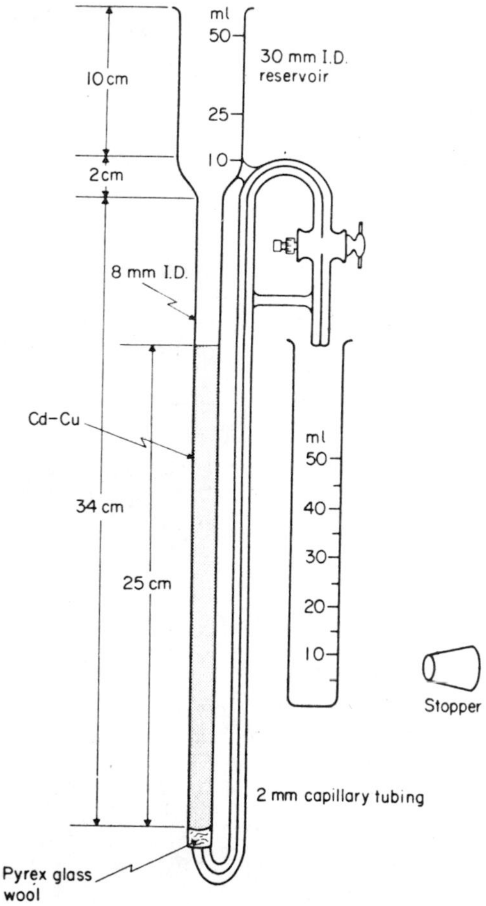

Figure 5.2. Column used to reduce nitrate to nitrite, with collecting cylinder (Wood, Armstrong and Richards, 1967). Teflon stopcock with metering valve (Kimax stopcock no. 41007-F, $1\frac{1}{2}$ mm bore and capillary arms). Figure reproduced with permission of Cambridge University Press.

Reagents

A KNO_3 standard solution, 2 μg ml^{-1} of NO_3–N

See § 5.1 solution (B).

B HCl, 2 M
Dilute 12 M HCl (see §1.3.4).

C Cupric sulphate, 2%
Dissolve 20 g $CuSO_4.5H_2O$ in 1 litre H_2O.

D Buffer solution
Dissolve 100 g NH_4Cl, 20 g $Na_2B_4O_7$ and 1 g of $Na_2EDTA.2H_2O$ in H_2O and make up to 1 litre.

E Cadmium metal filings, about 0·5 mm diameter
Different cadmium sources may give different results. See Interferences.

Apparatus
50ml stoppered graduated cylinders.
Cadmium–copper reduction columns (Fig. 5.2).

Prepare each column by washing 5 g cadmium filings with 25 ml 2 M HCl and rinsing with H_2O. Add 10 ml cupric sulphate solution (C) and swirl the contents until the blue colour disappears. Plug the bottom of the column with glass wool and fill with water. Add treated cadmium to a point level with the outlet. There must be no trapped air bubbles. Flush the column twice with a solution of 50 ml H_2O plus 5 ml buffer (D). Adjust the stopcock until the flow is 25 ml in 240 ± 10 secs.

Procedure
Prepare duplicate 50 ml aliquots of nitrate standard for each cadmium–copper column, of concentration approximating that of samples for analysis. Place 50 ml H_2O, standard, or sample in a stoppered graduated cylinder. Add 5 ml of buffer solution (D) and invert to mix.

Place 10 ml of buffered sample on the column; allow it to run through. Discard the effluent. Add the remainder of the buffered solution to the column and collect 2 ml effluent in the same cylinder; rinse and discard. Collect 25 ml of the column effluent. Run a H_2O blank and a standard through each column used. NO_2^- is determined in the column effluent as in § 5.4.1.

Precision and range
The operating range is 0–500 $\mu g\ l^{-1}$ of NO_3–N.

Precision is claimed to be about 1% (but results may become erratic).

Interferences
Different cadmium sources give different results. Standard addition is essential to check the efficiency of the reduction column.

Particulate matter should be removed by filtration. Sulphide and some organic materials (e.g. humic compounds) cause low results (Afghan and Ryan, 1975). These interferents can be removed by boiling with $K_2S_2O_8$ (see § 5.7.3) (aliquots can be used for NO_3^- and PO_4–P), or by passing through a column of chromatographic Al_2O_3. Standard addition must be used to check losses during these treatments.

LEVEL III

The cadmium–copper reduction method can be automated. For the reduction either a cadmium column or a coil of cadmium wire may be used.

5.4 NITRITE

LEVEL I AND II

5.4.1. Colorimetric

Principle

In a strongly acid medium HNO_2 reacts with sulphanilamide to form a diazonium compound, which reacts quantitatively with N-(1-naphthyl)ethylenediamine dihydrochloride to form a strongly coloured azo compound. If any NO_2^- is left at this stage it will destroy the reagent (D) (see below) so that almost no colour will develop, and the sample will *appear* to contain almost no NO_2^-. Possible excess NO_2^- is therefore destroyed by adding ammonium sulphamate ($NH_2SO_3NH_4$) just before reagent (D) (see below). This situation also allows the use of as near perfect a reagent blank as may be obtained; ammonium sulphamate is added to the water sample as first (rather than third) reagent. All NO_2^- is thus destroyed before colour development.

(Note that the upper limit of the range of the method is much smaller than the amount of NO_2^- stoicheiometrically equivalent to the added sulphanilamide.)

Reagents

A_1 KNO_2 standard, 0·05 M (0·7 mg ml^{-1} NO_2–N)

Dissolve 1·064 g of KNO_2 (A.R., dried at 105°C for one hour) in H_2O. Add 1 ml of 5 M NaOH and dilute to 250 ml. The solution can be stored, but must be checked oxidimetrically.

To do this, mix 10·0 ml of A_1 with 25·0 ml 0·1 M $\frac{1}{5}$ $KMnO_4$ (standardized), acidify with 5 ml 2 M H_2SO_4. Wait 15 minutes and add a known excess of $(COOH)_2$. Then backtitrate with the $KMnO_4$ solution.

The KNO_2 solution (A_1) contains 700 mg l^{-1} NO_2–N.

A_2 Diluted KNO_2 standard, 3·50 μg ml^{-1} NO_2–N

Dilute (A_1) 200 times (5 ml → 1000 ml). This solution contains 3·50 μg ml^{-1} NO_2–N.

B HCl, 6 M

Dilute 12 M HCl (see § 1.3.4).

C Sulphanilamide, 0·2%

Dissolve 2 g of sulphanilamide in 1 litre of H_2O.

D N-(1-naphthyl)ethylenediamine di-HCl, 0·1%

Dissolve 0·1 g of N-(1-naphthyl)ethylenediamine di-HCl in 100 ml of H_2O. The reagent can be kept for at least 2 months in a refrigerator and can be used even when it has become coloured. Store in a dark brown bottle.

E Ammonium sulphamate, 5%
Dissolve 5 g of NH_2SO_3 NH_4 in 100 ml of H_2O. Store at room temperature.

Storage
The concentration of NO_2–N can change very rapidly due to bacterial conversions, either by oxidation of NH_3, or by reduction of NO_3^-. When prolonged storage is necessary the samples can be mixed with reagents (C) and (B) in the field. The samples can then be kept for at least 24 hours if put in a refrigerator. Filtration can be carried out after development of the colour.

Procedure
Mix 100 ml of sample containing not more than 35 μg of NO_2–N (or an aliquot diluted to 100 ml) in a 100/110 ml volumetric flask with 5 ml of suphanilamide solution (C), and 2 ml of HCl (B). After 3 minutes add 1 ml of ammonium sulphamate solution (E) followed after another 3 minutes by 1 ml of naphthylethylenediamine solution (D). Dilute to 110 ml. If only 100 ml volumetric flasks are available use a 90 ml sample.

Prepare a blank by adding 1 ml of solution (E), and 2 ml HCl (B) before the sulphanilamide solution (C), is added. Measure the absorbance of the reagent blank, standards and samples in suitable cells at a wavelength as close as possible to 530 nm after 15 minutes.

Prepare a calibration curve in the range of 1–35 μg of NO_2–N using the diluted standard (A_2). In this range the calibration curve is linear.

Precision and accuracy
Precision and accuracy may be about 1½%.

.5 ORGANIC NITROGEN

LEVEL II

.5.1. Colorimetric

Dissolved organic nitrogen

Principle
Organic nitrogen (org.–N) is determined by a wet digestion of filtered samples either after the removal of the inorganic nitrogen, or as the difference between the value obtained for total nitrogen and that for ammonia (§ 5.1.1). Thus org.–N = [(org.-N) + (NH_3–N)] – (NH_3–N).

Removal of NH_3 and NO_3^- (after reduction) by a previous distillation (see 5.1.2) before the Kjeldahl destruction is suitable if the inorganic compounds are present in excess of the organic nitrogen compounds.

The less stable organic nitrogen compounds may hydrolyse to inorganic compounds (see 5.1.2), but the nitrogen in these organic compounds may be as easily available to algae as are NH_3 and NO_3^-. Biologically therefore the distinction may be unimportant.

The Kjeldahl digest can also be used for Total Phosphate determination (§ 5.7). The digest must then be divided into two portions, one for the NH_3 determination and one for the Total-P determination.

Sampling and storage

Filter the samples immediately after collection. Millipore filters contain organic nitrogen, but they can be used if the filtrates alone are analysed. Rinse the filters with H_2O before use. Blanks must also be filtered.

Because organic substances containing nitrogen may be continuously broken down to release NH_3, the only satisfactory methods of storage are to preserve with H_2SO_4 or to deep-freeze to $-20°C$ (after filtration). Direct investigation of fresh samples is preferable.

METHOD (A):

Digestion followed by distillation of the NH_3 produced.

Apparatus

The basic digestion vessel is the long-neck Kjeldahl flask (Fig. 5.3). For macro determinations the volume may be as large as 800 ml, for micro determinations about 30 ml. Complete multiple units are available, with gas or electrical heaters, H_2SO_4 fume chimney and distillation equipment. For routine work these are very satisfactory.

Reagents

H_2O See § 5.1

A H_2SO_4, (1 + 1) See § 1.3.4

Use the purest H_2SO_4 (A.R.) available. Some firms supply a 'nitrogen-free' grade. See note § 5.1.

B NaOH, 10 M

See § 1.3.4.

C $CuSO_4$ solution 10%

Dissolve 10 g of $CuSO_4.5H_2O$ (A.R.) in 100 ml of H_2O.

D K_2SO_4 (A.R.), solid (or Na_2SO_4 (A.R.))

E NaCl solution, 10%

Dissolve 10 g of NaCl (A.R.) in 100 ml of H_2O.

Procedure

Put not more than 40 ml of sample, preferably after procedure § 5.1.2, containing not more than 500 μg of nitrogen, in a 100 ml Kjeldahl flask (or smaller volume flask if only smaller samples are available) and mix with 4 ml H_2SO_4 (A), 10 drops (0·3 ml) of $CuSO_4$ solution (C)†, 3·0 g of K_2SO_4 (D)†, and 1 ml of NaCl (E)†. Add a quartz boiling chip or a glass bead. Heat the flask on a low flame for the first 10 to 30 minutes, until frothing stops. Then raise the temperature gradually until the sample is completely charred, and the acid boils and refluxes in one-third of the neck of the digestion flask. The flame should *not* be allowed to touch the flask above the part occupied by the liquid as this may lead to a loss of NH_3 because of the decomposition of $(NH_4)_2SO_4$.

†Not if after procedure § 5.1.2.

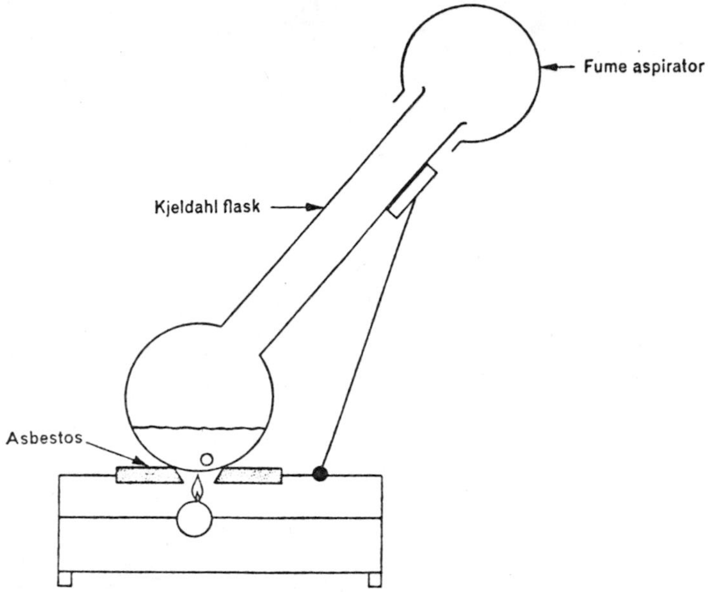

Figure 5.3. Kjeldahl digestion flask and simple digestion apparatus.

Avoid heating at a temperature greater than about 400°C, because of undue latilization of acid before all the organic matter is oxidized, and because some H_3 may be lost. Rotate the flask at intervals and continue heating until the organic atter is destroyed. This should happen about one hour after the solution has ared and has turned to a pale green colour. After digestion, allow the flask to ol. Dilute the contents with 10 ml of H_2O, and warm gently, if necessary, to solve the 'cake' of $KHSO_4$. Transfer the digest quantitatively to the Parnas–agner (or other design) distillation apparatus (Fig. 5.1) by rinsing three times h 5 ml of H_2O. Add 10 ml of 10 M NaOH (B) and close the tap A. Then follow ocedure § 5.1.1.

For macro determination *either* collect the distillate in an excess of H_3BO_3 in Erlenmeyer flask and then titrate with HCl or NH_2SO_3H (sulphamic acid, .2.2), *or* transfer an aliquot of the digest (containing not more than 500 μg of rogen) to the distillation apparatus. Neutralise the H_2SO_4 (A) with 10 ml 10 M OH (B) then distill off and determine the NH_3, as described in procedure .1.1.

cision, accuracy and range of application

ganic nitrogen content from 0·05 mg to 100 mg l^{-1} of N may be determined h an error of 0·05 mg l^{-1} N, the precision depending on the amount present, l on the accuracy of the apparatus and of the NH_3 determination.

tes

The distillation may be carried out directly from the Kjeldahl flask as shown in Fig. 5.4. The flask is heated (after the addition of excess NaOH) and the distillation is continued for two minutes after the steam has reached the receiving flask. The rate of heating should be adjusted so that the temperature

in the receiving flask is 80–90°C after two minutes of distillation. (Jönsson, 1966).

(b) Sudden temperature changes in the distillation apparatus (for instance when NaOH is added), or in the steam generator may result in liquid being sucked back.

(c) K_2SO_4 is added in order to raise the boiling point of the digest from 345 to 380°C, without too much loss of H_2SO_4. The temperature should not exceed 400°C or loss of nitrogen will result.

(d) The addition of NaCl is advisable for samples which have a low organic nitrogen and a relatively high nitrate content and which do not already contain Cl^-. The Cl^- will partly prevent the reduction of NO_3^- to NH_3, and will form products with NO_3^- (such as NOCl) which are to some extent volatile and do not interfere. The addition of NaCl is not necessary, if the NO_2^- and NO_3^- are removed after reduction as described in procedure § 5.1.

(e) All apparatus must be cleaned very thoroughly with acid dichromate cleaning mixture (§ 1.2.5), and blanks should be run in advance and regularly in between series of determinations.

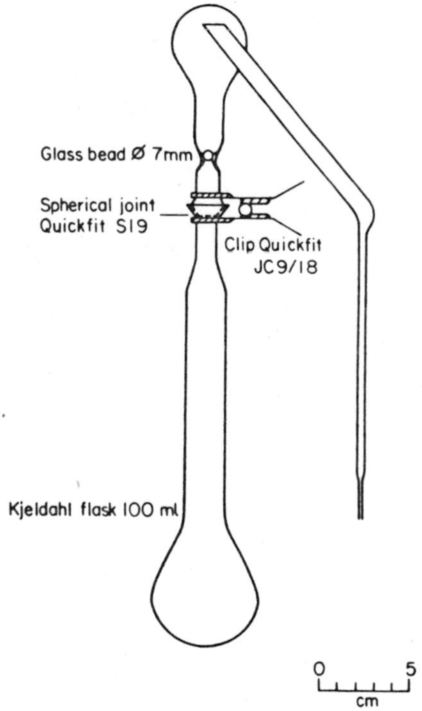

Figure 5.4. Micro-Kjeldahl distillation apparatus.

Method (B):

Digestion followed by determination of the NH_3 produced, directly in the neutralised digest (Scheiner, 1976).

Reagents

As § 5.2.5 Method (b), plus

F Digestive solution

Dissolve 134 g K_2SO_4 in 650 ml water, add 200 ml concentrated H_2SO_4 carefully with stirring. Add 5 ml selenyl chloride ($SeOCl_2$), dilute to 1 litre and mix.

Procedure

To 5 ml of sample in a Kjeldahl flask add 5 ml digestive solution (F) and mix. Insert a small glass funnel in the mouth of the flask and heat the flask gently under a hood. When the solution clears, boil briskly for some minutes and cool. (The digestion of 5 ml domestic wastewater is usually complete in 30 min.) Simultaneously treat 5 ml H_2O as a reagent blank. Transfer the contents of the flask to a beaker by washing with 35–36 ml 1 M NaOH. Raise the pH to 7 at least. Transfer the neutralised sample to a 100 ml volumetric flask and make up to the mark. Place 25 ml of diluted sample in a 50 ml volumetric flask and continue as in § 5.2.5, Method (b). Read the absorbance against the blank.

For N concentrations less than 0·4 mg l^{-1} larger samples should be used.

LEVEL II

5.5.2. Colorimetric

Particulate organic nitrogen

The method is in principle the same as for dissolved organic nitrogen. There are two variants:

(1) For samples containing large amounts of particulate organic nitrogen the Kjeldahl determination (with parallel NH_3 determination) can be carried out on an unfiltered and on a filtered sample. From the difference the particulate nitrogen can be calculated. Filtration through Millipore filters—after prerinsing—is satisfactory if blanks, filtered in the same way, are used.

(2) If the sample contains only small amounts of particulate organic nitrogen, it is better to concentrate the material on a glass fibre filter.

Samples with a high silt content should be filtered on a hard paper filter designed for quantitative work. Wash the filters with H_2O before use. Digest the samples with filters in a Kjeldahl flask. When using paper filters special attention should be given to the start of the digestion due to the large amount of material to be carbonized. Millipore filters are not suitable when particulate organic nitrogen content is small, due to their relatively high organic N content.

5.5.3. Deleted from this edition

Dissolved or total organic nitrogen

5.5.4. Colorimetric

Dissolved and particulate organic nitrogen

Organic nitrogen can be determined as NH_3 in the oxidation mixture of procedure § 7.3.2 or § 7.3.1 (Golterman, 1971). Partially neutralise the digest cautiously

—with half the theoretical amount of 10 M NaOH. Transfer the mixture to the distillation apparatus and neutralise completely with NaOH. Distill off the NH_3 and determine the amount as in § 5.1.1.

LEVEL III

5.5.5. UV oxidation

Total dissolved nitrogen

Total dissolved nitrogen may be determined after photo-oxidation by UV irradiation (Henriksen, 1970). Organo-nitrogen compounds and ammonia in solutions with pH range 6·5 to 9 are oxidised to NO_3^- and NO_2^- after 4 hr irradiation in the presence of excess O_2 (ensured by addition of a few drops 30% H_2O_2).

Apparatus

A photochemical reactor, consisting of a high pressure mercury–arc lamp, a reactor body with 24 fused silica sample tubes and a cooling fan.

Reagents

A $NaHCO_3$, 0·1 M

Dissolve 0·84 g of $NaHCO_3$ in 100 ml H_2O.

B H_2O_2, 30%

Procedure

Fill the silica tubes with about 40 ml well-mixed sample. Add 0·4 ml of 0·1 M $NaHCO_3$ and 2 drops 30% H_2O_2. Mix. Place the tubes in the reactor and irradiate for 4 hours. Cool and analyse for NO_3^- and NO_2^- on an Autoanalyser, or use procedures § 5.3.2 and § 5.4.1.

Precision

About 2% in the range 125 to 700 $\mu g\ l^{-1}$ of N.

5.5.6. C, H, N train

Commercial apparatus exists for the simultaneous determination of particulate C N (and H). If the concentration of these elements is sufficiently high the dissolved C, N (and H) can also be determined after evaporation.

PHOSPHORUS COMPOUNDS

Principle (see Olsen, 1967)

In strongly acid solutions ortho-phosphate (PO_4–P) will form a yellow complex with molybdate ions. This can then be reduced to a highly coloured blue complex. If ascorbic acid is used as a reducing agent, the formation of the blue colour is stimulated by antimony. The method can be made more sensitive by extraction of the blue complex into an organic layer, or by extraction of the yellow complex before reduction.

Phosphate bound to organic substances (org-P) and polyphosphate (poly-P) do not react with the molybdate reagent. These compounds must therefore be hydrolysed in order to convert the phosphate to H_3PO_4. High temperature and a high acidity are essential for the digestion.

Analysis of the hydrolysate of filtered water gives the quantity Total Dissolved Phosphate (Tot-P_{diss}). Subtraction of the PO_4-P then gives the Hydrolysable Phosphate (Poly-P + Org-P_{diss}). In unpolluted waters this fraction is usually organic phosphate only. In polluted waters the polyphosphate concentration may be high.

Analysis of the hydrolysate of the unfiltered sample gives Total-P. Subtraction of the Tot-P_{diss} then gives the Particulate Phosphate (Part-P). Particulate Phosphate can also be determined by hydrolysis of the filter residue, if the filters are, or are washed, phosphate free.

Sampling and storage

Any conventional sampler may be used, but the greatest care is needed to avoid contamination or loss by adsorption to the walls of containers. It is best to measure and put unfiltered samples directly from the sampler into the reaction vessel. For filtered samples, filter immediately after collection through a membrane filter, pore size about 0·5 μm. Prerinse the filter with at least 250 ml H_2O. Store the measured filtrate in the reaction vessel. The determination should be carried out as soon as possible.

If delay cannot be avoided the filtered sample may be stored by deep-freezing at −20°C, in a polyethylene bottle with a polyethylene stopper. Frozen samples should be thawed completely and mixed carefully before analysis.

Phosphorus may be released from living cells on freezing, or from dying cells. This can increase the proportion of dissolved phosphorus if unfiltered samples are frozen, melted and later filtered. The total amount of phosphorus should not change.

If glass bottles must be used, Pyrex or Jenaglass (borosilicate) is usually satisfactory. Ordinary soda-lime glass must be avoided.

5.6 ORTHO-PHOSPHATE

LEVEL I

5.6.1. Colorimetric

Reagents

A KH_2PO_4 standard, 40 μg ml^{-1} of PO_4–P See § 5.6.2

B H_2SO_4, 2 M See § 1.3.4

C Reagent mixture

Mix the following dry reagents thoroughly: 48 g of $(NH_4)_6Mo_7O_{24}.4H_2O$, 1·0 g of sodium antimony tartrate and 40 g of ascorbic acid. This mixture is, if kept dry, stable for three months.

D Mixed reagent solution

Dissolve 1 ± 0·2 g (a 'constant spoonful') of the reagent mixture (C) in 100 ml of 2 M H_2SO_4. Allow to stand for about 15 minutes before using. This reagent should not be kept for more than one day (Fishman and Skougstad, 1965).

Procedure

Mix 40·0 ml of sample and 10·0 ml of reagent (D) in an Erlenmeyer flask. Measure the absorbance of the reagent blank, standards and samples in suitable cells in a simple colorimeter after at least 10 minutes. Alternatively match the colour against a phosphate standard.

If necessary the modifications (b) and (c) of method § 5.6.2 can be used to make the determination more sensitive.

Precision and accuracy

The precision and accuracy are estimated at 5–10% depending on the colorimeter available.

Interferences

See method at level II.

LEVEL II

5.6.2. Colorimetric

Reagents

A KH_2PO_4 standard solution, 40 μg ml^{-1} of PO_4–P

Dissolve 0·1757 g of KH_2PO_4 (dry, A.R.) in H_2O and dilute to 1000 ml. This solution contains 40 μg P per ml. Prepare a 40 fold dilution (25·0 ml → 1000 ml), which contains 1 μg of P per ml. Add 2·5 ml of H_2SO_4 (1 + 1) and a few drops of $CHCl_3$ as a preservative to both standard solutions before making up to volume.

B H_2SO_4, 2 M See § 1.3.4

C Molybdate-antimony solution

Dissolve 4·8 g of $(NH_4)_6Mo_7O_{24}.4H_2O$ and 0·1 g of sodium antimony tartrate ($NaSbOC_4H_4O_6$) in 400 ml of 2 M H_2SO_4 (B) and make up to 500 ml with the same acid (Murphy and Riley, 1958 and 1962).

D Ascorbic acid (about 0·1 M)

Dissolve 2·0 g of ascorbic acid in 100 ml of H_2O. This solution is usable for one week if kept in a refrigerator.

E n-Hexanol (reagent grade)

F Isopropanol (reagent grade)

Procedure

Modification a

Pipette 40·0 ml of the sample into a 50 ml volumetric flask. Add 5 ml of molybdate (C) and 2 ml of ascorbic acid solution (D) and mix well. Dilute to 50 ml. Measure the absorbance of the reagent blank, standards and samples in suitable cells at a wavelength as close as possible to 882 nm (or with a filter with maximum light transmittance at 720–750 nm) at least 10 minutes after mixing.

Modification b (Stephens, 1963)
If the absorbance is below 0·100 but above 0·020 (using 5 cm cells) transfer the coloured sample quantitatively into a separatory funnel. Add 10 ml n-hexanol and shake vigorously for 1 minute. Discard the lower layer and transfer the hexanol to a 10 ml graduated cylinder. Adjust the volume to 10 ml with a few drops of isopropanol, which clears the solution. Transfer to cells which have at least as long a light path as those used in modification **a**, but which are nevertheless sufficiently filled by 10 ml. Measure the absorbance of the reagent blank, standards and samples in suitable cells, at a wavelength as close as possible to 690 nm (or with a filter with a maximum wavelength transmission at 720 nm).

Modification c
If the absorbance is still too low, larger volumes of the sample can be extracted, e.g. 100 ml or 200 ml. In these cases add 10 or 20 ml of solution (C) and 4 or 8 ml of solution (D). The extraction can still be carried out with only 10 ml n-hexanol. Make up to 10 ml with isopropanol. About 1 ml should be needed.

Prepare a calibration curve which covers the range 50–500 $\mu g\ l^{-1}$ of PO_4–P for modification **a**, 10–100 $\mu g\ l^{-1}$ for modification **b**, and 2–20 $\mu g\ l^{-1}$ for modification **c**.

Precision and accuracy
The precision for modification **a** is probably $1\frac{1}{2}$%; for the others 2–5%, depending on the care with which the extraction is carried out. Extraction is 90–95% complete, the proportion extracted varying with temperature and solvent/solution ratio. The accuracy depends on the calibration curve for the extraction methods. The best results will be obtained by standard addition. The minimal requirement is that standards be run with each series. The calibration curve for this method is linear up to about 0·5 mg l^{-1} of PO_4–P. If necessary, the linear range may be extended to about 3 mg l^{-1} (Harwood, van Steenderen and Kühn, 1969).

Interferences
Cu^{2+}, Fe and silicate (up to 10 mg l^{-1} of Si) do not interfere. Arsenate produces a colour similar to that produced by phosphate and can be removed (see § 5.7.3.) Humic acids and oxidizable organic material interfere and must be removed, using the method § 5.7.3.

5.6.3. Colorimetric (tentative)

Principle
Extraction of the yellow phosphomolybdate in isobutanol and reduction with $SnCl_2$ to the blue compound. (Proctor and Hood, 1954; Golterman and Würtz, 1961.)

Reagents

A **KH_2PO_4 standard** See § 5.6.2

B **H_2SO_4, 2 M** See § 1.3.4

C Molybdate solution, 5%
Dissolve 25 g of $(NH_4)_6Mo_7O_{24}.4H_2O$ in 500 ml of 2 M H_2SO_4.

D $SnCl_2$ solution, 0·1%
Dissolve 2 g of hydrazine sulphate (anti-oxidant) in 1 litre of 0·3 M H_2SO_4. Cool to about 10°C and add 1 g of $SnCl_2.2H_2O$. This solution becomes clear after about 12 hours in a refrigerator and can be kept at 4°C for 4–8 weeks at least.

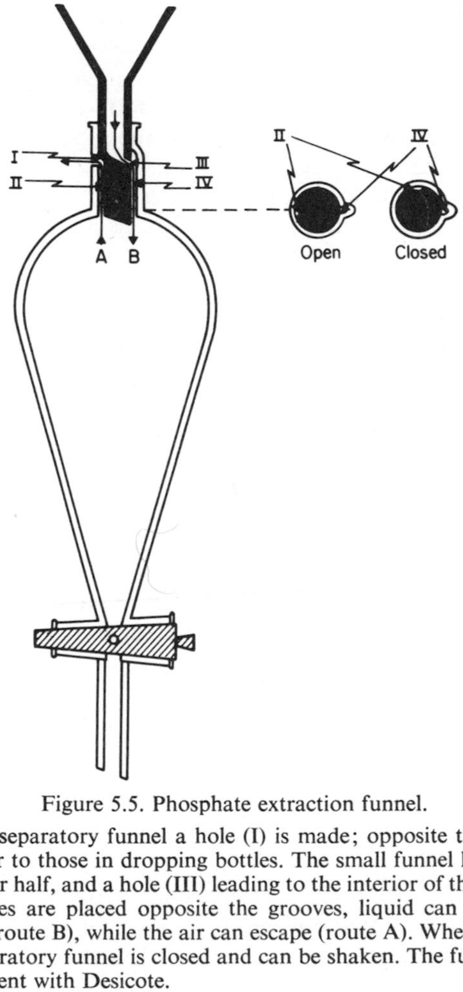

Figure 5.5. Phosphate extraction funnel.

In the neck of the separatory funnel a hole (I) is made; opposite this hole lies a small groove (IV), similar to those in dropping bottles. The small funnel has a similar groove (II) in its solid lower half, and a hole (III) leading to the interior of the separatory funnel. When the two holes are placed opposite the grooves, liquid can be poured into the separatory funnel (route B), while the air can escape (route A). When the small funnel is turned 90° the separatory funnel is closed and can be shaken. The funnel is made water-repellent by treatment with Desicote.

E Isobutanol
Use A.R. grade or distill the laboratory reagent quality. Use the fraction 107–108·5°C.

F Ethanol 96% (normal quality)

G $\frac{1}{5}KMnO_4$, 0·02 M
Dissolve 3 g of $KMnO_4$ in 1 litre of H_2O.

Apparatus

The apparatus shown in Fig. 5.5 is convenient for the extraction technique.

Procedure

Mix 50 ml of sample in a separatory funnel with 6·0 ml of 2 M H_2SO_4 and with 5 ml of molybdate reagent (C).

The yellow complex with phosphate is formed, though it may not be apparent if the phosphate concentration is low.

Add 15 ml of isobutanol (E) and shake for 15 sec. Discard the aqueous layer and wash (shake) with 5 ml of 2 M H_2SO_4 (B) plus 10–20 ml of H_2O. Remove the aqueous layer quantitatively. (Allow the organic layer to fill the bore of the stopcock of the funnel.) Dry the funnel outlet. Rinse the organic layer with ethanol into a suitable volumetric flask (25 or 50 ml) and add 0·25 ml (for 25 ml flask) or 0·50 ml (for 50 ml flask) of $SnCl_2$ solution (D). Dilute to volume with ethanol. Measure the absorbance against a reagent blank in suitable cells at about 720 nm within 30 minutes. If the blue colour is too weak a larger sample can be used with proportionally increased quantities of H_2SO_4 and molybdate reagent. 15 ml more isobutanol (E) must be added than is necessary for saturation (about 10 ml per 100 ml sample).

If desired, 10–20 ml of n-hexanol may be substituted for iso-butanol.

Precision and accuracy

A precision and accuracy of 1–2% may be obtained.

Interferences

When the water has a high C.O.D. difficulties arise. These can be avoided by adding 7 ml 2 M H_2SO_4 to the sample, followed by sufficient $KMnO_4$ (G) for the colour to remain for at least 15 minutes. (This can be done in the separatory funnel.) Remove the excess of $KMnO_4$ by adding isobutanol and shaking. Then add the acid molybdate reagent. When large amounts of humic acids are present it may be possible to remove them by the procedure just described. They can alternatively be removed by extraction of the acidified sample with isobutanol, until the organic layer is no longer coloured. The organic layers can be collected and evaporated to dryness with 1 ml of 2 M H_2SO_4. The phosphates that were extracted with humic acids will dissolve in the H_2SO_4. (It is not known whether or not phosphates are partly split off from the humic acids during the extraction.) The phosphates can be determined after destruction of these humic acids (see § 5.7).

When silicates are present at concentrations less than 0·05 mg l^{-1} the concentration of the acid during the extraction may be changed to 0·2 M. Add no extra acid to the sample before the molybdate reagent.

LEVEL III

5.6.4. By AAFS

Great sensitivity can be obtained by measuring the molybdate in the organic solvent by atomic absorption flame spectrophotometry.

5.7 TOTAL PHOSPHATE

LEVEL II

5.7.1. Hydrolysis with H_2SO_4 digestion

Principle

Polyphosphates and some organic phosphorus compounds are hydrolysed by H_2SO_4 to orthophosphate, which is then determined by the method of § 5.6.2 or § 5.6.3. The addition of H_2O_2 completes the destruction of organic material.

Reagents

A H_2SO_4 (1 + 1) See § 1.3.4

B $CuSO_4$ solution, 10%
Dissolve 10 g $CuSO_4.5H_2O$ in 100 ml H_2O.

C H_2O_2, 30% A.R.
Check that a PO_4-free grade is used.

Special apparatus

Use special Kjeldahl flasks, Fig. 5.3, (about 75 ml) with a neck about 30 cm long. These prevent the loss of P_2O_5 by volatilization. All glassware must be cleaned with hot chromic sulphuric acid. New glassware should then be filled with slightly acidic H_2O and left to stand for some weeks. Note that most detergents and laboratory cleaning solutions contain phosphorus and must be avoided.

Procedure

Put 50 ml of sample (or, if desired, a smaller volume) in the long necked Kjeldahl flask. Add 4 ml H_2SO_4 (A) and 10 drops (0·5 ml) of the $CuSO_4$ solution (B). Evaporate—if necessary, add more sample—and heat to fuming. If much organic matter is present and black charred material is produced, cool, and add 10 drops of H_2O_2 (C). Heat until a clear solution is obtained and continue heating for 15 minutes. It is necessary to remove all the H_2O_2 before adding molybdate. To do this, cool and add about 5 ml of H_2O. Then heat to fuming to remove the last traces of H_2O_2. Cool, add 20 ml of H_2O, and boil.

If phosphate is to be determined by procedure § 5.6.2 the digest must be nearly neutralized with NaOH and then diluted to 100 ml (or smaller volume if desired). Take 40 ml for the phosphate determination. A second aliquot may be used for the determination of organic nitrogen, § 5.5.1. If procedure § 5.6.3 is used the neutralization is not necessary.

5.7.2. Method deleted from this edition

5.7.3. Hydrolysis with H_2SO_4

Principle

Polyphosphates and organic phosphorus compounds are hydrolysed by H_2SO_4 and $K_2S_2O_8$ to orthophosphate, which is then determined by the method of § 5.6.2 or § 5.6.3. This is a suitable method if many samples are to be analysed.

Procedure

Use Erlenmeyer flasks. Add sufficient H_2SO_4 to make the sample 0·15 M acid. For rapid digestion or if much organic matter is present, add 1 g $K_2S_2O_8$ (PO_4-free). Cover the flask with an inverted beaker and heat in an autoclave (or pressure cooker) at 120°C for 2 hours if $K_2S_2O_8$ has been added, for 8 hours if it has not.

Procedure § 5.6.3 (PO_4–P) can then be used directly, omitting the 6 ml of 2 M H_2SO_4 and shaking with isobutanol for at least 1 minute.

Alternatively use procedure § 5.6.2 (PO_4–P) but with the molybdate–antimony reagent (C) made in 0·8 M H_2SO_4.

Interferences

Arsenate forms a blue complex with molybdate. This interference can be removed by reduction of the arsenate to arsenite (a) by the addition of a few drops of 5% KI solution, or (b) by the addition of thiosulphate in an acidic medium (Johnson, 1971).

LEVEL III

5.7.4. UV oxidation

Total P may be determined after photo-oxidation by UV light in the presence of acid (Henrikson, 1970). Organic compounds of P are converted to orthophosphate by irradiation for 1 hour in the presence of acid and excess O_2 (ensured by the addition of a few drops of 30% H_2O_2).

Apparatus

As § 5.5.5.

Reagents

A H_2SO_4, 4 M

Add 22 ml H_2SO_4 (s.g. 1·84) cautiously with stirring to H_2O and dilute to 100 ml.

B H_2O_2, 30% See § 5.7.1.

Procedure

Fill each silica tube with 40 ml well-mixed sample. Add 0·4 ml 4 M H_2SO_4 and 2 drops 30% H_2O_2. Mix. Irradiate the tubes for $1\frac{1}{2}$ hours. Cool and analyse for orthophosphate on an Autoanalyser, or use procedures § 5.6.2 or § 5.6.3.

Precision

The precision is about 1% in the concentration range 4 to 240 $\mu g\ l^{-1}$ P.

SILICA COMPOUNDS

The formulae of the compounds of silicon which are determined by these methods are not known. The terms 'silica' and 'silicate' are used here as a convenient description only. (See Golterman, 1967b).

5.8 MOLYBDATE REACTIVE SILICA

LEVEL II

5.8.1. Colorimetric

Principle

Between pH 3 and 4 silicate forms a yellow complex with molybdate ions which can then be reduced to a highly coloured blue complex.

Reagents

A Na_2SiF_6, standard solution* (100 μg ml^{-1} of SiO_3–Si)

Dissolve 671·4 mg of Na_2SiF_6 (A.R.) in 400 ml of H_2O, heating until solution is complete. Cool and dilute to 1000 ml. Store in a thick walled polythene bottle. This solution contains 100 μg ml^{-1} of Si.

A_1

Dilute solution (A) 10 times (10·0 ml → 100 ml). This solution contains 10 μg ml^{-1} of Si.

A_2

Dilute solution (A) 50 times (10·0 ml → 500 ml). This solution contains 2 μg ml^{-1} of Si.

B H_2SO_4, 1 + 1 See § 1.3.4

C Na_2MoO_4 solution, 5%

Dissolve 5 g of Na_2MoO_4 in 100 ml of 0·25 M H_2SO_4. Allow to stand for 48 hours before use.

D_1 $SnCl_2$, (stock solution)

Dissolve 40 g of $SnCl_2$ in 12 M HCl (s.g. 1·18) at room temperature and dilute to 100 ml with 12 M HCl. Allow to stand for 24 hours before use.

D_2 Diluted $SnCl_2$ solution

Dilute 1 ml of (D_1) with 100 ml of H_2O. Use immediately.

Procedure

Mix 20 ml of sample (pH 5–8) containing not more than 100 μg Si, (or an aliquot diluted to 20 ml) with 2 ml of solution (C). Allow to stand for 15 minutes, and

*Na_2SiF_6 may be obtained from Riedel de Haën or Baker Co.

add 5 ml of H_2SO_4 (B). Cool to room temperature and add 1 ml of the freshly prepared solution (D_2). Measure the absorbance of the reagent blank, standards and samples in suitable cells at a wavelength as close as possible to 815 nm 10–15 minutes after mixing. A calibration curve—covering the range 5–100 μg of Si per sample—must be made. Run a blank using 20 ml of H_2O.

Tentative procedure
In waters with a very high silicate concentration it is possible that colloidal silicate dissolves after dilution (see Kobayashi, 1967). In this case dilution must be avoided and the yellow colour should be measured after the addition of the H_2SO_4. Measure the absorbance at 365 nm. As this is not the maximum of the absorption curve (see Golterman, 1967b) the wavelength must be checked carefully, and the wavelength adjustment must be free from drift; a filter colorimeter is therefore actually preferable to most spectrophotometers.

Precision, accuracy and range of application
The precision is estimated at 1–2%. With a good calibration curve the precision and accuracy are equal.

The range of application is 0·1–5·0 mg l^{-1} SiO_3–Si for the blue method. The yellow method can be used in the range of 1·0–20·0 mg l^{-1} of Si.

5.9 TOTAL SILICA

LEVEL II

5.9.1. Colorimetric

Principle
Silica may occur in the water in colloidal and in particulate form as well as in solution. That which passes through a 0·5 μm filter may be considered as soluble or colloidal. The particulate fraction is that retained by the filter. Particulate silica is made soluble by fusion with $NaKCO_3$ in a platinum crucible.

The silica is converted into sodium silicate and is then determined as described below. Particulate silica may be either determined directly, or obtained by difference between total and soluble plus colloidal silica.

Reagents

A H_2SO_4, 2 M See § 1.3.4

B $NaKCO_3$, 2%
Dissolve 20 g of $NaKCO_3$ (A.R.) in 1 litre of H_2O. Store in a polythene bottle.

Procedure
Evaporate a sample, containing not more than 100 μg of Si, to dryness together with 10·0 ml of $NaKCO_3$ (B) in a platinum crucible. Gently fuse the dry residue using a small flame at first and then a Meker burner for 5 minutes. An electric

furnace at a temperature of 900°C is preferable. Cool. Dissolve the melt in 40 ml of H_2O, and transfer to a 100 ml polythene bottle. Rinse the crucible twice and add the rinsings. Add 1 ml of H_2SO_4 (A) slowly to the polythene bottle. After mixing, dilute the solution to volume in a 100 ml volumetric flask.

Avoid contamination from the outside of the crucible. (Platinum crucibles in an electric furnace are best placed on a platinum disc.)

Then proceed as described for reactive silicate, § 5.8.

Interferences

(For reactive silicate and total silica procedures.)

The similar yellow and blue complexes with phosphate interfere, but are destroyed by the H_2SO_4 added. H_2SO_4 is more efficient than oxalic acid. When the phosphate concentrations are high a control is necessary. (Run a silicate blank, containing the same amount of phosphate as the sample, against a normal blank.)

In the yellow method self-colour and turbidity of the sample may interfere.

Keep reagents and samples in thick walled polythene flasks. Keep the samples in the dark. With some waters deep-freezing has led to errors due to precipitation of the silica (Kobayashi, 1967).

CHAPTER 6

TRACE ELEMENTS

A BORON, COBALT, COPPER, IRON, MANGANESE, MOLYBDENUM, VANADIUM, ZINC, AND ALUMINIUM

B CADMIUM, LEAD AND MERCURY

LEVEL II (colorimetric) or III (AAFS)

6.0 *General*

In dealing with trace elements a useful distinction can be made between two groups:

A, those elements which occur naturally in water and most of which are plant nutrients and **B**, those elements potentially toxic in low concentrations and which have been widely distributed as a result of human activities. The elements of group **A** will normally be measured to obtain insight into the factors controlling biological characteristics, while the elements in group **B** are often monitored for health reasons.

Copper and zinc are examples of elements which may be in both groups: they occur naturally, are essential plant nutrients, but the present concern is mainly related to their possible toxicity. Nevertheless this distinction is useful because the source of the elements of the first group is general erosion and in studying an unpolluted lake the limnologist has to consider the importance of these elements. Their sources are diffuse and consequently many samples may be necessary to get adequate estimates. The second group, of pollution elements, is supplied more often from point sources, e.g. industrial or household sewage.

In many countries the law requires that such sources be notified by the body responsible for their production.

Because of their low concentrations, the analysis of trace elements presents acute problems, and only procedures at level II or III are possible. In particular, contamination or losses (as a result of adsorption on samplers for example) can occur very easily. Dust in the laboratory is a particular problem of prime concern, because this interference is never reproducible in blanks. For reliable results standard addition should be used, together with H_2O blanks (to check reagent blanks) as described in § 1.2.3. Blanks should be taken through all stages including extraction, filtration and digestion. It is indeed advisable to take H_2O into the field and transfer it to a sample bottle there in the same way that the lake water samples are.

The validity of each method for a particular situation should be checked. Trace elements (especially the heavy metals) normally occur in several forms: dissolved, 'ionic', 'chelated' (the most abundant form), 'colloidal' (for example basic copper carbonate) and 'suspended matter' (either adsorbed on silt or as heavy metal oxides).

It is therefore useful to make analyses of samples which have been:

1. filtered
2. filtered and digested
3. not filtered
4. not filtered, but digested.

For the determination of dissolved constituents the sample should be filtered through a 0·45 μm pore membrane filter as soon as possible. Use the first 50 ml to rinse the filter flask. Discard this portion and collect the filtrate, which should be acidified as soon as possible. In studies of *total* loading the samples should be digested without filtration, but care should be taken to ensure that particulate matter remains homogeneously dispersed.

Because trace elements can occur in so many states and at such low concentrations it is even more difficult than usual to interpret analyses in terms of 'availability' to organisms. Bioassay may be the only solution to this problem, but bioassays have a complex of attendant difficulties that put them beyond our scope (see Golterman, 1975 p 194).

There are many techniques for determining the concentration of trace elements. The most important ones are colorimetry (often after a concentration step), polarography (classical, cathode ray and anodic stripping voltammetry) and atomic absorption flame spectrophotometry, AAFS (Fabricand *et al*, 1962; Fishman and Downs, 1966).

The advantage of colorimetric methods is that in some cases a distinction between different states (for example 'ionic' and 'chelated') may be made. If a concentration step is needed however then this advantage no longer exists. Concentration may be carried out by extraction or by use of ion exchangers.

Solvent extraction, which is also an important concentration step before AAFS, can generally be made by chelating the metals with APDC (ammonium pyrrolidine dithiocarbamate) and extracting with MIBK (methyl-iso-butylketone). The general procedure using APDC and MIBK is described in § 6.0.3. Other extraction procedures (e.g. with dithizone) are described with the individual methods.

The polarographic methods, of which anodic stripping voltammetry is the most sensitive, were once supposed to be specific in measuring ionic species only. Davison (personal communication) considers that both d.c. polarography and differential pulse polarography measure the electroactive component and any other component that is in sufficiently rapid equilibrium within the time scale of the measurement.

For example, nickel ions in an aqueous-ammonia system exist in the following states:

$$Ni(H_2O)_6^{2+} \rightleftharpoons Ni(NH_3)(H_2O)_5^{2+} \rightleftharpoons Ni(NH_3)_2(H_2O)_4^{2+} \rightleftharpoons Ni(NH_3)_6^{2+}$$

It has been demonstrated that only one of these seven species, $Ni(NH_3)_2(H_2O)_4^{2+}$, is electroactive but because the equilibria are rapid the current measured is due to the concentration of this species plus the concentration of the other species.

If we have Fe^{2+} in H_2O then the reducible species may be considered to be $Fe(H_2O)_6^{2+}$. If EDTA is added almost all the ferrous iron is present as $Fe(EDTA)^{2+}$, but we have the equilibrium $Fe(H_2O)_6^{2+} + EDTA \rightleftharpoons FeEDTA^{2+} + 6\ H_2O$, which is governed by the appropriate stability constant K. If we measure this solution polarographically we know the $Fe(H_2O)_6^{2+}$ will be reduced and the rate of exchange of the Fe^{2+} between the H_2O and the EDTA ligands is very rapid, so although $Fe(H_2O)_6^{2+}$ is in very small concentration in solution it appears that we are measuring Fe $EDTA^{2+}$. The size of the signal for the $Fe^{2+} - H_2O$ system and the Fe^{2+} – EDTA system will be the same, but the kinetic influence will have the effect of shifting the potential of reduction slightly.

The difference between the d.c. polarographic measurement and the differential pulse measurement rests entirely in the time scale of the two techniques. The d.c. technique makes a measurement in 0·5 to 5 seconds, that is, the drop frequency, but the pulse technique measures in no more than 60 milliseconds after the application of a pertubation. So the pulse technique will exclude some species with slow equilibria which were included in the d.c. measurement.

In anodic stripping voltammetry the solution is pre-electrolysed at a fixed potential to concentrate the deposited metal as an amalgam in the mercury drop.

Because this process takes as long as 3 minutes, the metal deposited will depend on the convective diffusion transport in solution (e.g. stirring rate). The electrode kinetics themselves would not be expected to control the eventual current signal. In common with the other two techniques, however, any non reducible chemically inert (commonly referred to as strongly bound) complexes will not be measured. For further details see Davison (1976 and 1978).

The quickest, and amongst the most sensitive, technique is **AAFS**, which is now widely used. No fractionation of species can be made. The sensitivity can be high. Some indication of the analytical characteristics of one of the best machines is given in table 6.1. The range given for the sensitivity depends on the make of the apparatus.

Table 6.1. Characteristics for several trace elements of one of the best atomic absorption flame spectrophotometers.

	Sensitivity (detection limit)* with or without solvent extraction. [mg l⁻¹]		concentration for 1% absorption [mg l⁻¹]	best working range of concentration [mg l⁻¹]
	with extraction	without extraction		
Al	0·05 –1·0	0·1 –1·0	0·4	10 –1000
Cd	0·001 –0·01	0·001–0·01	0·004	0·1– 2
Co	0·001 –0·01	0·01		0·1– 10
Cr	0·0002–0·01	0·01	0·02	1·0– 200
Cu	0·001 –0·01	0·005–0·01	0·04	0·1– 10
Fe	0·001 –0·05	0·004–0·05	0·006	0·1– 20
Pb	0·001 –0·05	0·01 –0·05	0·06	1·0– 10
Mn	0·001 –0·01	0·005–0·01	0·04	0·1– 20
Zn	0·001 –0·01	0·005–0·01	0·02	0·1– 2
Mo	0·0002–0·05	0·05		0·1– 10

* Defined here as twice the fluctuation in background signal.

Details of AAFS are given in § 1.4.3. For organic solvents a special burner may be needed and the fuel to air ratio should be reduced because the organic solvent contributes to the fuel supply. For some elements a nitrous oxide-acetylene flame (and special burner) is required. Special grades of HCl and HNO_3 for AAFS for trace metals are commercially available. Otherwise distillation from all-glass apparatus should be used to purify these reagents.

Fishman and Erdmann (1975) reviewed new techniques and reactions for trace metals in water. In some cases the procedures are still being tested and are not yet standardised.

6.0.1 Sampling and storage

Samples, even those of surface waters, must be collected from below the surface film. Completely non-metallic samplers must be used and lowering cables or ropes should be of nylon or similar materials or coated with plastic. Even rubber should be avoided as it contains significant amounts of zinc. Samplers may be tested as sources of contamination by storing samples in them and removing aliquots for analysis at 15 minute intervals.

Containers used for field collection should be of polyethylene with stoppers of polyethylene or a similar acid resistant plastic. They must be washed before use with concentrated HCl or 50% HNO_3 and rinsed with H_2O. They may profitably be rinsed with sample just before filling. Analyses should be performed in borosilicate glass vessels which have been cleaned with strong acid. Running a blank of H_2O which has been stored in a representative manner is recommended. Long term storage is not recommended. If it is necessary to store samples for more than a few hours it should be done in the vessels in which final analysis will be made, i.e. samples for digestion should be stored in the digestion flasks, and the samples to be filtered should be filtered and stored in the flasks in which they will be analyzed. This will reduce problems caused by precipitation and adsorption. If gross storage for several days is unavoidable the samples should be filtered (if appropriate) and about 5 ml of concentrated HNO_3 added to each litre. (This renders a later distinction between ferric and ferrous iron impossible.) Further digestion may be needed if, for example, clays are present. In some cases the HNO_3 can be evaporated and the residue can be dissolved in 1 M HCl. If much clay is present the samples must be digested with hydrofluoric acid at 105°C in a decomposition vessel consisting of a stainless-steel container fitted with a removable PTFE (Teflon) crucible with screw cap.* Hendel (1973) used these vessels for the analysis of glass with an acid dissolution with HF and boric acid. Agemian and Chau (1975) used a mixture of nitric, perchloric and hydrofluoric acids (3·5 hours at 140°C) for all kinds of sediments including sand, and organic-rich silt. In the digest they were able to measure 20 different elements among which were Cd, Cr, Co, Cu, Mg, Pb, Zn, Mo, V, Mn, Fe, Al, Ba, Ca. Krishnamurty *et al* (1976) used a (safe) HNO_3–H_2O_2 mixture for digestion. A preliminary pre-oxidation of the samples with HNO_3 eliminated the danger of uncontrollable reactions.

6.0.2 Sample digestion

Several methods are available, including digestion with HNO_3, $HClO_4$ and $K_2S_2O_8$. An ultra-violet photo-oxidation has been described by Armstrong *et al* (1966), Armstrong and Tibbits (1968) and Henriksen (1970). The apparatus is described

* Expensive but satisfactory versions from: Uni-seal Decomposition Vessels Ltd., P.O. Box 9463, Haifa, Israel.

in § 5.5.5. Because the $HClO_4$ procedure is potentially dangerous (Everett 1967) it is omitted here. If the limnochemist insists on using $HClO_4$ he should mix it with other acids, (and ensure that his affairs are in order).

Persulphate digestion is convenient and rapid (Hansen, and Robinson, 1953; Menzel and Corwin, 1965). Add a freshly prepared 5% solution of $K_2S_2O_8$ to the sample in the ratio of 8 to 50 (V/V). Heat the sample mixture in a waterbath at 100°C for 60 minutes or autoclave (or pressure cook) at 120°C for 30 minutes. Use normal Erlenmeyer flasks covered with a suitable beaker. The digest is strongly acid because of decomposition products of the persulphate.

For the determination of suspended solids transfer the filter (§ 6.0) to a beaker and add 3 ml distilled (or very pure) HNO_3. Cover the beaker with a watch glass and heat.

6.0.3 Solvent extraction

If the concentration of a metal is not sufficiently high for a direct determination then a concentration step must be used. Chelation and extraction are the usual methods. Ammonium pyrrolidine dithiocarbamate (APDC) can be used for Cu, Cd, hexavalent Cr, Fe, Mn, Pb and Zn, but Cr^{3+} must first be oxidised to the hexavalent form with $KMnO_4$. Manganese can only be extracted if the pH is adjusted to 4·5 to 5·0. The Mn-complex is unstable and must be analysed without delay. The most frequently used solvent for APDC is methyl isobutyl ketone (MIBK). It provides a stable flame in AAFS and its physical properties such as viscosity, surface tension, boiling point and mutual solubility in an aqueous solution are favourable. It gives a low background in the flame.

Reagents

A APDC solution, 1%

Dissolve 1 g of ammonium pyrrolidine dithiocarbamate in 100 ml of H_2O.

B Methyl isobutyl ketone, special grade, MIBK

(If strongly acid solutions must be extracted it is advisable to mix 3 volumes of MIBK with 1 volume of cyclohexane).

Extraction procedure

Transfer 100 ml to a 250 ml container (tapering beaker or wide mouthed 'Griffin' flask) with a standard interchangeable stopper (or use a smaller aliquot diluted to 100 ml). Adjust the pH of the sample to 2·5 with HCl or HNO_3 using a pH meter. Add 2·5 ml of solution **A** and 10 ml of solution **B**, stopper and shake the flask vigorously for one minute. Allow the two layers to separate and transfer the organic layer to a narrow long necked container, for example a measuring cylinder or volumetric flask of 25 ml. If necessary add H_2O to raise the organic layer into the upper part of the container and aspirate the ketone directly into the AAFS.

6.1 BORON

Colorimetric

Principle

Boron forms a coloured complex with either carmine, resorcinol, or curcumin. The carmine complex method is given here, and is taken from Standard Methods

(1965). The resorcinol complex method (Glebovich, 1963) also seems suited to routine analysis. The curcumin method (Greenhalgh and Riley, 1962) is probably the most accurate and is suggested for critical work.

Reagents

A Standard Boron solution, 100 $\mu g\ ml^{-1}$
Dissolve 0·5716 g of H_3BO_3 in H_2O and dilute to 1000 ml; 1·00 ml contains 100 μg B. (Because H_3BO_3 loses H_2O on drying at 105°C a pure reagent should be used and kept tightly stoppered to prevent absorption of moisture).

B H_2SO_4 (A.R., s.g. 1·84) See § 1.3.4.

C HCl, 12 M (A.R.) See § 1.3.4.

D HCl, 1 M
Dilute 4 M HCl (see § 1.3.4) 0·25 litre → 1 litre.

E NaOH, 1 M
Dilute 10 M NaOH (see § 1.3.4) 0·1 litre → 1 litre.

F Carmine solution, 0·1%
Dissolve 1 g of carmine (NF 40, or carminic acid, the main constituent of carmine) in 1 litre of H_2SO_4 (s.g. 1·84). This may take up to 2 hours.

6.1.1 Boric acid

Procedure
Put 2·00 ml sample into a small Erlenmeyer flask or 30 ml test tube, and add 2 drops (0·1 ml) of HCl (C) and, with great care, 10·0 ml of H_2SO_4 (B). Mix well and cool. Add 10·0 ml of carmine solution (F), mix well and allow to stand for at least 45 minutes. Measure the absorbance of the reagent blank, standards and samples in suitable cells at a wavelength as close as possible to 585 nm.

Prepare a calibration curve corresponding to 2–20 μg of Boron per sample.
N.B. When the liquids are mixed, bubbles form. The mixing should therefore be efficient, and bubbles be allowed to dissipate before measuring the absorbance.

6.1.2 Total Boron

Procedure
Make a suitable aliquot just alkaline with 1 M NaOH (E), and then add a slight excess of NaOH. Prepare a blank containing the same amount of NaOH. Evaporate both sample and blank to dryness on a steam bath. Ignite at 500–550°C. Cool and add 2·5 ml of 1 M HCl (D). Make sure the solution is acid, and mix with a rubber tipped glass rod. Pour the solution into a conical centrifuge tube and centrifuge it until clear. Pipette 2·00 ml of the clear solution into an Erlenmeyer flask and proceed as above.

Precision, accuracy, and sensitivity
Results are estimated to be accurate and reproducible within $\pm$ 0·4 μg of B per sample. The sensitivity is 0·2 μg of B per sample. If higher sensitivity is required the curcumin method is recommended.

6.1.3 Automated method

Automated methods for boron have been described by Lionnel (1970) and Afghan *et al* (1972).

6.2 COBALT

Colorimetric

Principle

2-nitroso-l-naphthol yields a coloured complex with cobalt. The complex can be extracted into non-polar solvents. The excess reagent is removed from the organic phase with NaOH. The following procedure is modified from Sandell (1959) and Bilikova (1973).

Reagents

A_1 Standard Co solution, 100 μg ml^{-1}

Dissolve 0·0404 g of $CoCl_2.6H_2O$ in water, add 3 ml of 4 M HCl, dilute to 100 ml. This solution contains 100 μg ml^{-1} of Co.

A_2 Standard Co solution, 10 μg ml^{-1}

Dilute A_1 ten times with 0·1 M HCl

B HCl, 2 M

Dilute 4 M HCl (see § 1.3.4); (50 ml $\rightarrow$ 100 ml).

C NaOH, 2 M

Dilute 10 M NaOH (see § 1.3.4); (20 ml $\rightarrow$ 100 ml).

D Sodium citrate, 40%

Dissolve 200 g of sodium citrate in H_2O and dilute to about 500 ml.

E H_2O_2, 3%

Dilute H_2O_2 30% (A.R.); (10 ml $\rightarrow$ 100 ml).

F 2-nitroso-l-naphthol solution, 1%

Dissolve 1·0 g of 2-nitroso-l-naphthol in 100 ml of glacial CH_3COOH. Add 1 g of activated carbon. Shake the solution before use and filter off the required amount. (This procedure keeps the reagent blank low).

G $CHCl_3$ or 1, 2-dichloroethane

Procedure

Add 10 ml of sodium citrate solution (D) to 100–500 ml (exactly measured) of sample. Adjust the pH to 3–4 with 2 M HCl (B) or 2 M NaOH (C). Add 10 ml of 3% H_2O_2 (E), mix, and allow to stand for a short while. Then add 2 ml of reagent solution (F), mix, and allow to stand for at least 30 minutes at room temperature. Transfer to a separatory funnel and shake the funnel vigorously for 1 minute with 25 ml of $CHCl_3$ (G). Separate the solvent. Repeat the extraction twice with 10 ml portions of $CHCl_3$ (G). Dilute the combined extracts to 50 ml and transfer to a clean separatory funnel. Shake the funnel for 1 minute with 20 ml 2 M HCl (B), and then run the $CHCl_3$ layer into another funnel and shake it for 1 minute with 20 ml 2 M NaOH (C). Finally, separate the $CHCl_3$ layer for absorbance measurement. Measure the absorbance of the reagent blank, standards and samples in suitable cells at a wavelength as close as possible to 530 nm. The colour is stable for at least 12 hours. Prepare a calibration curve in the range 5–200 μg of Co, dissolved in the same volume of H_2O as the samples.

Interferences

If the sample contains more than 2 mg l^{-1} of Fe it should be pretreated with sufficient 2 M HCl to dissolve the Fe. Fe^{2+} interferes, but in this procedure it is oxidized by the H_2O_2 and kept from precipitation and extraction by the citrate. Ni^{2+} and Cu^{2+} could interfere but are prevented from doing so by the HCl wash of the $CHCl_3$ extract. Unless the sample contains enough Ni^{2+} and Cu^{2+} to cause problems this step could be left out. The NaOH wash is still necessary.

Precision, accuracy, and sensitivity

Precision and accuracy are unknown. 200 μg of Co (in a final volume of 50 ml of $CHCl_3$) in a 1 cm cell have an absorbance of about 1·0 (Sandell, 1959). Thus it is possible to determine as little as 1–2 μg.

6.2.2 Atomic absorption flame spectrophotometric

Cobalt may be measured by AAFS. There are no serious interferences. Wavelength 240·7 nm. Oxidising air-acetylene flame.

The sensitivity is rarely sufficient for natural waters, but may be improved 10 fold by using solvent extraction with APDC plus MIBK (see § 6.0.3).

6.3 COPPER

6.3.1 Colorimetric

Principle

Copper, reduced to its cuprous form, reacts with diquinolyl to form a coloured complex, which is extracted into an organic solvent for colorimetric estimation. (Riley and Sinhaseni, 1958.)

Reagents

A Standard Cu solution, 1 mg ml^{-1}

Weigh accurately 100 mg of Cu (A.R. or electrolytic) in a quartz or Pyrex beaker. Dissolve it in 3 ml of HNO_3 (s.g. 1·42), add 1 ml of H_2SO_4 (s.g. 1·84) and evaporate under an infrared heater until dense white fumes are evolved. Allow to cool and then dissolve the residue in H_2O and dilute to 100 ml.

A_1 Standard Cu solution, 10 μg ml^{-1}

Dilute solution (A) 100 times (10·0 ml → 1000 ml). Solution (A_1) contains 10 μg ml^{-1} of Cu.

A_2 Standard Cu solution, 2 μg ml^{-1}
Dilute solution (A_1) 5 times (10·0 ml → 50·0 ml). Solution (A_2) contains 2 μg ml^{-1} of Cu. Prepare (A_1) and (A_2) afresh as required.

B $NH_2OH.HCl$ solution, 10%
Dissolve 10 g of $NH_2OH.HCl$ (A.R.) in about 80 ml of H_2O, filter and dilute to 100 ml. If the reagent contains appreciable amounts of Cu^{2+}, extract it with 10 ml portions of a 0·01% solution of dithizone in CCl_4 (see § 6.8) until there is no change in the colour of the dithizone. Extract solution (B) with CCl_4 until all colour has been removed.

C Sodium acetate buffer solution, 1 M
Dissolve 136 g of $CH_3COONa.3H_2O$ in H_2O and dilute to 1 litre. If the reagent contains more than a trace of Cu, extract with dithizone as described above.

D Diquinolyl reagent, 0·03%
Dissolve 0·03 g of 2, 2′-diquinolyl in 100 ml of n-hexanol that has been redistilled over NaOH.

E Ethanolic hydroquinone solution, 1%
Dissolve 1 g of hydroquinone in 100 ml of redistilled ethanol.

Procedure
The analyses are made in 1 litre separatory funnels. Clean these by leaving them overnight filled with a (1 + 1) mixture of HNO_3 (s.g. 1·42) and H_2SO_4 (s.g. 1·84). Empty them, and rinse several times with H_2O.

Transfer 900 ml of sample to a 1 litre separatory funnel and add 10 ml of $NH_2OH.HCl$ solution (B) and 10 ml of buffer solution (C) (to adjust pH to 4·3–5·8). Add 8 ml of diquinolyl reagent (D) and shake the funnel for 5 minutes. Run the aqueous layer into another funnel, add to it 2 ml $NH_2OH.HCl$ solution (B) (to keep the Cu reduced), and 3 ml diquinolyl reagent (D). Extract again for 3 minutes. Repeat this extraction once more.

Combine the three hexanol extracts in a calibrated cylinder containing 0·5 ml hydroquinone solution (E). Dilute to 15 ml with hexanol. Measure the absorbance of the reagent blank, standards and samples in suitable cells at a wavelength as close as possible to 540 nm. The colour is stable.

N.B. Hexanol is recommended by Riley and Sinhaseni (1958), but a 50/50 mixture of CCl_4 and isoamyl alcohol may be preferred because it is heavier than water. Reagent (D) may be made in this solvent mixture, and there is then no need to transfer the sample to another funnel for each extraction.

Interferences
None are likely in natural waters.

Precision and sensitivity
The method gave a dispersion of 2·5% with seawater containing 27·0 μg l^{-1} of Cu. Samples ranging from 2 to 100 μg per 500 ml sample yielded a mean absorbance of 0·0393 per μg of Cu^{2+}.

6.3.2 Atomic absorption flame spectrophotometric
Copper may be measured by AAFS. There are no serious interferences.

Solvent extraction with APDC plus MIBK (§ 6.0.3) may be used for greater sensitivity.

Wavelength 324·7 nm. Oxidising air-acetylene flame.

6.4 IRON

Colorimetric

Principle

Iron, after reduction to Fe^{2+}, reacts with bathophenanthroline to form a red compound which, if necessary, is extracted into hexanol and estimated colorimetrically (Lee and Stumm, 1960).

Iron can exist as Fe^{2+} or Fe^{3+}. The generally accepted approach is to determine total inorganic iron and Fe^{2+}, and to obtain Fe^{3+} by difference. Total iron, including organic Fe, may be estimated after a wet digestion.

For estimation of 'available iron', see Shapiro (1967b).

Reagents

A_1 Standard Fe^{2+} solution, 10 μg ml^{-1}

Dissolve 0·0702 g $(NH_4)_2SO_4.FeSO_4.6H_2O$ (Mohr's salt), in about 0·5 litre H_2O that contains 5 ml H_2SO_4 (1 + 1), and dilute to 1000 ml. This solution contains 10 μg ml^{-1} Fe^{2+}. Standardize the Mohr's salt with 0·100 M $\frac{1}{5}$ $KMnO_4$.

A_2 Standard Fe^{2+} solution, 1·0 μg ml^{-1}

Dilute 10 ml of solution (A_1) to 100·0 ml with H_2O. This solution contains 1·0 μg ml^{-1} of Fe^{2+}, and should be prepared afresh every day.

B Standard Fe^{3+} solution, 10 μg ml^{-1}

Add excess aqueous chlorine to an aliquot of 10 μg ml^{-1} standard Fe^{2+} solution. Boil to remove unreacted aqueous chlorine. Dilute the cool solution to the original aliquot volume.

C HCl, 4 M See § 1.3.4.

D NH_4OH, 4 M

Dilute NH_4OH (s.g. 0·91) 0·3 litre → 1·0 litre.

E CH_3COONa, 10%

Dissolve 10 g of CH_3COONa (or 15 g of $CH_3COONa.3H_2O$) in 100 ml of H_2O contained in a separatory funnel. To remove traces of iron, add 2 ml 0·001 M bathophenanthroline (G), and mix well. Add 10 ml n-hexanol, and extract. Repeat the extraction in a second separatory funnel to ensure complete removal of iron. Store the solution in a glass stoppered bottle.

F $NH_2OH.HCl$, 10%

Dissolve 10 g of $NH_2OH.HCl$ in 100 ml H_2O. Remove traces of iron by the same procedure as that described for reagent (E). The solution has a pH of 1·5–1·75. Store in a glass stoppered bottle.

G Bathophenanthroline solution, 0·001 M
Dissolve 0·0332 g of 4,7-diphenyl-l,10-phenanthroline ($C_{24}H_{16}N_2$, mol wt 332) in 50 ml of ethanol, and dilute to 100 ml with iron-free water. Store the solution in a glass stoppered bottle.

H n-Hexanol
Reagent grade n-hexanol may be used without further purification. Technical grade material must be distilled before use.

J Ethanol, 95%
Re-distill commercial grade ethyl alcohol to remove iron.

Procedure

6.4.1 Total Iron (inorganic or, if following digestion, inorganic plus organic)
Add 5 ml of 4 M HCl (C) and 2 ml of 10% $NH_2OH.HCl$ (F) solution to a 50 ml sample, or to a neutral digested sample diluted to 50 ml. Add a glass bead, and boil for 15 minutes, cool, and make up to approximately the initial volume with H_2O. Add 5 ml of bathophenanthroline (G), mix, and add 4 ml 10% CH_3COONa solution (E). Adjust the pH to 4–5 with 4 M NH_4OH (D) added drop by drop. This point may be determined with pH indicator paper (Congo red), or as the point where the solution becomes cloudy. Transfer the solution quantitatively to a 125 ml separatory funnel, add 10·0 ml n-hexanol (H), and shake the funnel for 3 minutes. Discard the aqueous layer. Run the hexanol solution into a 50 ml cylinder, rinse the funnel contents into this with 95% ethanol (J) from a squeeze bottle, and make up to 50 ml with the ethanol. Measure the absorbance of the reagent blank, standards and samples in suitable cells at a wavelength as close as possible to 533 nm within 10 minutes.

6.4.2 Fe^{2+}
Proceed as for Total Iron (§ 6.4.1), but do *not* add $NH_2OH.HCl$ (F), and do *not* boil. Do not let the sample stand in the acidified condition but carry the procedure through immediately. If it is not possible to continue at once, the sample may be stored for several hours in a full, stoppered bottle but the acid should not be added until the procedure can be completed.

Interferences
Lee and Stumm (1960), from whose method this procedure has been modified, state that Co and Cu (both of which can form yellow complexes with bathophenanthroline) do not interfere. $HClO_4$ in high concentrations causes precipitation of bathophenanthroline. If this is a problem in the particular circumstances then the persulphate digestion method should be used. Fe^{3+} interferes in the Fe^{2+} determination, giving a small amount of colour with the reagents. Its contribution can be estimated from a curve prepared by making the Fe^{2+} analysis on a series of standards containing only Fe^{3+}. Lee and Stumm suggested in their original method that the sample be boiled with acid (p. 1572, § 3.3). If any organic matter is present however, Fe^{3+} will be converted to Fe^{2+} (Shapiro, 1965; O'Connor *et al* 1965), and this step is therefore best omitted for Fe^{2+} determinations.

Sensitivity
Using a 4 cm cell 1 μg of Fe^{2+} per litre of sample gives an absorbance of 0·001. If desired, the hexanol may be diluted less for greater sensitivity.

6.4.3 Atomic absorption flame spectrophotometric

Iron may be measured by AAFS. There are no serious interferences. Solvent extraction with APDC plus MIBK (§ 6.0.3) may be used for greater sensitivity.

Wavelength 248·3 nm. Oxidising air-acetylene flame.

6.5 MANGANESE

Colorimetric

Principle

Mn is oxidized to MnO_4^- which is then estimated colorimetrically. Either KIO_4 or $K_2S_2O_8$ may be the oxidizing agent. The persulphate method, which is described here, is simpler, as the sample need not be evaporated to remove chloride. The procedure is from Sandell (1959) and Standard Methods (1965). See Morgan and Stumm (1965) for a discussion of methods.

Manganese can exist in natural waters as soluble Mn^{2+} which is oxidizable to MnO_4^- by $K_2S_2O_8$ and also as insoluble higher valence forms such as MnO_2 which are not oxidizable to permanganate. Filtration does not necessarily separate these forms in natural waters, so determination of manganese may become rather complicated. Oxidation of organic matter by wet ashing will also leave the manganese in its oxidized state. Separate procedures are therefore suggested; one for soluble manganous ion (§ 6.5.1), one for total inorganic manganese (§ 6.5.2), and one for total manganese (§ 6.5.3).

Reagents

A Standard Mn solution, 50 mg l^{-1}

Dissolve 3·2 g of $KMnO_4$ in H_2O and make up to 1 litre. Heat for several hours near the boiling point, then filter through a sintered glass filter and standardize against $\frac{1}{2}(COOH)_2$ of accurately known concentration, approximately 0·1 M. From this $\frac{1}{5}KMnO_4$ solution prepare a standard containing 50 mg l^{-1} of Mn. To 1 litre of this add 5 ml H_2SO_4 (1 + 1) (see § 1.3.4) and then 10% $NaHSO_3$ solution in single drops, with stirring, until the permanganate colour disappears. Boil to remove the excess SO_2, and cool. Dilute again to 1000 ml with H_2O.

B H_3PO_4 (s.g. 1·69)

C $NH_2OH.HCl$ solution, 10%

Dissolve 10 g of $NH_2OH.HCl$ in 100 ml of H_2O.

Solution D_1 is alternative to D_2

D_1 Hydroquinone solution, 1%

Dissolve 1 g of hydroquinone in 100 ml of H_2O.

D_2 H_2O_2, 30%

E Special Hg/Ag solution

Dissolve 75 g of $HgSO_4$ in 400 ml of HNO_3 (s.g. 1·42) and 200 ml of H_2O. Add 200 ml of H_3PO_4 (s.g. 1·69), and 35 mg of $AgNO_3$. Dilute the cooled solution to 1 litre.

F $K_2S_2O_8$ (A.R.) Dry.

Procedure

6.5.1 Soluble Mn^{2+}. Filtered or unfiltered samples may be used.
Add 5 ml of solution (E) and 3 g of $K_2S_2O_8$ (F) to 90 ml of the sample, or to an aliquot diluted to 90 ml. Bring to boiling in about 2 minutes over a flame (do not heat on a water bath). Remove the flask from the flame, allow to stand for one minute and then cool under the tap. Dilute to 100 ml with H_2O.

Measure the absorbance of the reagent blank, standards and samples in suitable cells at a wavelength near 525 or 545 nm. The colour is stable if the persulphate is in excess. If the solution is cloudy during measurement add 1 drop of 1% hydroquinone solution (D_1) or 1 drop of 30% H_2O_2 (D_2). Stir and measure the absorbance again. The difference is the absorbance due to permanganate.

6.5.2 Total inorganic manganese. Filtered or unfiltered samples may be used.
Acidify a 90 ml sample with 3 ml H_3PO_4 (B). Add 2 drops $NH_2OH.HCl$ solution (C), and heat to boiling. Cool and proceed as in § 6.5.1.

6.5.3 Total manganese
Following wet digestion, proceed as in § 6.5.1.

Interferences
Fe^{2+} could interfere but in this procedure it is kept from doing so by the H_3PO_4. Interference by Cl^- is prevented by the special solution (E) (up to 100 mg of NaCl per sample).

Precision, accuracy, and sensitivity
The error in the determination of 0.05–5 mg of manganese per sample in the absence of interfering substances is usually not more than 1%. Method § 6.5.1 probably measures some organic Mn^{2+} if it exists, due to oxidation of some of the organic matter by the persulphate. The chief problem with this method is its low sensitivity: 1 mg l^{-1}. For this reason a more sensitive method (Strickland and Parsons, 1968) is also given here, which may be suitable. It may be used in place of the previous method for all Mn fractions.

6.5.4 Manganese (tentative)

Principle
Mn^{2+} ions are oxidized to MnO_4^- by KIO_4 in a sodium acetate buffer. The $KMnO_4$ is then used to oxidize leucomalachite green and the absorbance measured at about 620 nm (Strickland and Parsons, 1968). This method may be used to determine other Mn fractions following the filtration or digestion described in § 6.5.1–6.5.3.

Reagents

A Standard manganese solution, 0·5 μg ml^{-1}
See § 6.5, but dilute 100 fold just before use.

B Acetate buffer, 0·5 M

Dilute 30 ml CH_3COOH, glacial (A.R.), to 1000 ml with H_2O. Add carefully tc this 5 M NaOH solution (A.R.) until the pH of the buffer lies in the range 4·1 to 4·2 Store the solution in a thick walled polyethylene bottle. The solution is stable for many months.

C KIO_4 solution, 0.2%

Dissolve 1·0 g of KIO_4 (A.R.) in 500 ml of H_2O, and add one small pellet (abou 0·2 g) of NaOH. Store the solution in a dark glass bottle out of direct sunlight The solution is stable for several weeks if stored in the dark, but slowly decompose when exposed to strong light. For accurate work the solution should be made up every few days.

D Leucomalachite Green solution, 0·08%

Dissolve 0·20 g of leucomalachite green (4,4-Bisdimethylaminotriphenylmethane which has been recrystallized from alcohol in 250 ml of pure acetone. This solution is stable but should not be allowed to evaporate. The development of a slight green colour does not affect the results.

Procedure

Put 30 ml of sample into a 50 ml stoppered graduated measuring cylinder. Add acetate buffer (B) from a polyethylene wash bottle, to make the total volume in the cylinder 45 ml. Place the cylinder in a water bath at a temperature between 23 and 26°C, with a thermostat with accuracy of ± 1°C. After the solution has reached the bath temperature (15 to 30 minutes) add 5·0 ml of KIO_4 solution (C) from a pipette. Mix the solution thoroughly and leave in the bath for a further 10 to 1: minutes. Then add with rapid mixing, 1·0 ml of solution (D) from a pipette and return the cylinder to the thermostatically controlled bath.

Between 4 and 5 hours after adding the solution (D) measure the absorbance o the reagent blank, standards, and samples, in suitable cells at a wavelength a close as possible to 615 nm.

Interferences

Iron interferes, but its absorbance is very low compared to that resulting from th same molar concentration of Mn. The absorbance for a given Mn concentration increases as the concentration of salts decreases. Standard addition is therefor essential.

Precision, accuracy, and sensitivity

According to Strickland and Parsons (1968), as little as 0·1 $\mu g\ l^{-1}$ can be detected Precision and accuracy are unknown.

6.5.5 Atomic absorption flame spectrophotometric

Manganese may be measured by AAFS. Solvent extraction can be made with 1% oxine (8-hydroxyquinoline) in MIBK at pH between 9 and 11 (adjusted with NH_4OH).

Wavelength 279·5 nm. Oxidising air-acetylene flame.

Smith (1974) used an ion-exchange technique for the concentration of manganese from seawater.

6.6 MOLYBDENUM

6.6.1 Colorimetric

Principle
Molybdenum is co-precipitated with MnO, and determined colorimetrically with KCNS. The procedure of Bachmann and Goldman (1964) has proved satisfactory. An alternative procedure by Chan and Riley (1966b) may be more sensitive, but is more complicated.

Reagents

A_1 Standard Mo solution, 100 μg ml^{-1}
Dissolve 0·0750 g of MoO_3 (A.R.) in 10 ml of 0·1 M NaOH. Dilute with H_2O to about 50 ml, make slightly acidic with HCl, and make up to 500 ml with H_2O. 1 ml contains 100 μg of Mo.

A_2 Standard Mo solution, 10 μg ml^{-1}
Dilute A_1 tenfold with 0·1 M HCl. Prepare fresh solution each week.

B Buffered $MnSO_4$ solution
Dissolve 1·0 g of $MnSO_4.H_2O$ in 150 ml of H_2O. Mix 30 ml of this solution with 120 ml of glacial acetic acid and 50 ml of 4 M sodium acetate (544 g l^{-1} of $CH_3COONa.3H_2O$).

C $KMnO_4$ solution, 1%
Dissolve 1 g of $KMnO_4$ in 100 ml of H_2O.

D $SnCl_2$ solution, 10%
Dissolve 10 g of $SnCl_2.2H_2O$ in 100 ml of 1 M HCl. Filter if cloudy. Prepare fresh solution each week.

E KCNS reagent

E_1 HCl, 1·3 M
Dilute 4 M HCl (§ 1.3.4) three times (100 ml → 0·3 litre).

E_2 $(NH_4)_2SO_4.FeSO_4$ solution, 1%.
Dissolve 1 g of $(NH_4)_2SO_4.FeSO_4.6H_2O$ in 100 ml of 0·1 M H_2SO_4. Prepare fresh solution each day.

E_3 KCNS solution, 10%
Dissolve 10 g of KCNS in 100 ml of H_2O.

Mix 160 ml of 1·3 M HCl (E_1) with 10 ml of $(NH_4)_2SO_4.FeSO_4$ solution (E_2) and 30 ml of the KCNS solution (E_3). *Prepare fresh solution each day.*

F Solvent mixture of isoamyl alcohol and CCl_4
Mix equal volumes of A.R. products.

G $CaCl_2$ dry, (A.R.)

Procedure

Put 1 litre of sample in a 2 litre flask and heat to boiling. Remove source of heat, and add with vigorous stirring, 10 g of $CaCl_2$ (G), 20 ml of the buffered $MnSO_4$ solution (B), then 2 ml of the $KMnO_4$ solution (C). Stir occasionally until the hydrated $Mn(OH)_2$ forms and coagulates. Allow to cool to room temperature, collect the precipitate on a filter such as a Whatman GF/B, or a 47 mm membrane filter (ca. 1·2 μm pore size). At this point the precipitate may be stored for later analysis. Place the filter flat on the bottom of a 150 ml beaker or roll it and place it in a test tube. Add 5 ml of the KCNS reagent (E). Mix, by swirling, until the precipitate dissolves and transfer the solution to a 60 ml separatory funnel. Wash the filter with three additional 5 ml portions of the KCNS reagent (E), adding each to the separatory funnel. To the funnel add 3·0 ml of the $SnCl_2$ (D) and mix. Add exactly 5·0 ml of the solvent mixture (F) and shake the funnel vigorously for $1\frac{1}{2}$ minutes. Allow the aqueous and organic phases to separate. Separate the organic phase by filtering the bottom layer through a hard filter paper (e.g. Whatman no. 541) directly into the spectrophotometer cells. The filter paper removes water droplets and any insoluble matter which may be present. Wait about 10 minutes for any trace of pink, due to $Fe(CNS)_3$, to disappear.

Measure the absorbance of the reagent blank, standards and samples in suitable cells at a wavelength as close as possible to 465 nm. The colour is stable.

Interferences

The only interference likely to be present in natural waters is caused by iron, which produces a pink colour due to $Fe(CNS)_3$. After 10 minutes the iron reduces, and the interference disappears. In cases where much iron is present, wash the organic layer with $SnCl_2$ solution (D).

Precision, accuracy, and sensitivity

Recovery of added molybdenum averaged 99% for concentrations of 2·0–10·0 $\mu g\ l^{-1}$. Standard deviation was 2%.

6.6.2 Atomic absorption flame spectrophotometric

Molybdenum may be determined by AAFS after solvent extraction. Normally levels are so low that concentration is essential. To 100 ml water sample add 3 ml 1% benzoin-oxine in ethanol. Adjust the pH to 1·6 with HNO_3, then extract with n-butylacetate. Wavelength 313·3 nm. Reducing nitrous oxide-acetylene flame. With a graphite furnace (§ 1.4.3) it is possible to determine molybdenum in as little as 50 ml water in the $\mu g\ l^{-1}$ range after a selective collection of molybdenum with polymers at pH 2·5 (Muzzarelli and Rochetti, 1973).

6.7 VANADIUM

There is insufficient interest, at present, in the occurrence of V in fresh waters to justify the inclusion of a detailed method in this manual. Such methods may be found in Chan and Riley, (1966a), Fishman and Skougstad (1964) and in Fishman and Erdmann (1975).

6.8 ZINC

6.8.1 Colorimetric

Principle

Zinc reacts with dithizone in weakly acidic medium to form a red compound, which is extracted with CCl_4 and estimated colorimetrically (O'Connor and Renn, 1963).

Reagents

A Standard Zn solution, 1 mg ml^{-1}

Dissolve 1·00 g of granulated Zn (A.R.) in a slight excess of 4 M HCl (about 10 ml). Make up to 1 litre with H_2O. 1 ml contains 1 mg of Zn. Dilute standards should be freshly prepared.

B HCl, 4 M

Dilute 12 M HCl (§ 1.3.4) three times (1 + 2). Select a supply with a low blank, or redistill a 50% solution.

C Acetate buffer, 4 M

Mix equal volumes of 4 M CH_3COONa (544 g of $CH_3COONa.3H_2O$ in H_2O, dilute to 1 litre) and 4 M acetic acid (240 ml of $CH_3COOH \rightarrow$ 1 litre). Extract with dithizone solution (E) and subsequently with CCl_4 to remove zinc and colour. Store in dithizone-cleaned polyethylene bottles.

D $Na_2S_2O_3$ solution, 50%

Dissolve 500 g of $Na_2S_2O_3.5H_2O$ in H_2O and dilute to 1 litre. Remove zinc as described for reagent (C).

E Dithizone solution, 0·01%

Dissolve 10 mg of dithizone in 100 ml of CCl_4. ($CHCl_3$ may be used instead). Keep in refrigerator.

E_1 Dithizone solution, 0·001%

Dilute solution (E) 10 times (10 ml $\rightarrow$ 100 ml). Solution E_1 is usable for a few days only.

Procedure

Put 100 ml of the sample into a 125 ml separatory funnel and add HCl (B) to about 0·05 M (0·5–1·0 ml of 4 M HCl). Add 4 ml of buffer solution (C), 4 ml of $Na_2S_2O_3$ solution (D) and 10 ml of dithizone (E_1). Shake the funnel vigorously for 2 minutes. Separate the solvent and repeat the extraction with fresh dithizone. Combine the extracts and make up to a known volume with CCl_4. Measure the absorbance of the reagent blank, standards and samples in suitable cells at a wavelength between 520–540 nm (this measures the red zinc dithizonate) or at 620 nm (this measures the unreacted dithizone). Protect the solution from strong light.

Interferences

The most likely interferences in fresh water are due to Pb and Cu. Their reaction with dithizone is prevented by the $Na_2S_2O_3$ solution. If much Co or Ni is present it may be necessary to use KCN as a complexing agent. This is not usually necessary.

Precision, accuracy, and sensitivity
No figures available, but sensitivity is high. 1–2 μg of Zn per 100 ml of sample can be detected with ease.

6.8.2 Atomic absorption flame spectrophotometric
Zinc may be measured by AAFS with or without solvent extraction in MIBK (§ 6.0.3).

Wavelength 213·9 nm. Oxidising air-propane flame (direct aspiration) or air-acetylene flame (solvent extraction).

6.9 ALUMINIUM

6.9.1 Colorimetric

Principle
Aluminium forms a red complex with Aluminon (aurintricarboxylic acid) or Eriochrome Cyanine R. The colour formed depends on temperature, time of development and pH, all three of which must be standardised.

A more reliable, but tedious, method uses Ferron (8-hydroxy-7-iodoquinoline-5-sulphonic acid). The complex formed absorbs UV light. Iron would interfere but is chelated with ortho-phenanthroline. The iron may thus be determined simultaneously. Beryllium is added to minimise interference of (unnaturally high concentrations of) fluoride.

Reagents

A_1 Standard aluminium solution, 100 mg l^{-1}
Dissolve 1·758 g of $K_2Al_2(SO_4)_4.24\ H_2O$ in H_2O and dilute to 1000 ml; 1·00 ml contains 100 μg of Al.

A_2 Standard aluminium solution, 2 mg l^{-1}
Prepare daily a 50 fold dilution (20·0 ml → 1000 ml), which contains 2 μg ml^{-1} of Al.

B CH_3COONa, 35%
Dissolve 350 g of CH_3COONa (or 580 g of $CH_3COONa.3H_2O$) in H_2O and dilute to 1 litre.

C $NH_2OH.HCl$, 10%
Dissolve 100 g of $NH_2OH.HCl$ in H_2O. Add 40 ml of 12 M HCl (see § 1.3.4) and 1 g of $BeSO_4.2H_2O$ and dilute to 1 litre.

D Ferron-orthophenanthroline solution,
Add 0·5 g ferron and 1·0 g of ortho-phenanthroline to 1 litre of H_2O. Shake for several hours or heat mildly; allow any solids to settle out and decant the clear supernatant.

E Iron standard solution
See § 6.4, solution A_2.

Procedure

Put two portions of 25 ml of sample, one of H_2O and several standards of aluminium and of iron in separate beakers. Add 2·0 ml of solution (C) to each and allow o stand for 30 minutes. Add 5·0 ml of solution (D) to one sample, to H_2O and to he standards. Add 5·0 ml of H_2O to the other sample ('colour correction sample'). Add 2 ml of solution (B) to each. Between 10 and 30 minutes after adding solution B) measure the absorbance of sample, blank, standards and colour correction sample at 370 nm (for Al) and at 520 nm (for Fe) against H_2O. Subtract blank and colour correction from the measured sample absorbance, but blank alone from tandards.

Calculation

Correct the absorbance at 370 nm for the contribution from iron by calculating he iron concentration from the calibration curve at 520 nm (at which wavelength he Al complex has negligible absorbance). Calculate the absorbance of this con-entration of Fe at 370 nm and subtract it from the 'corrected absorbance' at 70 nm. Use this figure and the Al standard curve at 370 nm to estimate Al con-entration.

nterferences

Possible fluoride interference is suppressed by the beryllium in reagent (C). Mn interference, which occurs at concentrations above 1·0 mg l^{-1}, may be corrected n the same way as iron interference.

.9.2 Atomic absorption flame spectrophotometric

The best working range is 10–1000 mg l^{-1}, but concentrations as low as 1 mg l^{-1} nay be estimated by direct aspiration. Maintain an acid strength of 0·15% HNO_3 n all samples and standards. Sensitivity may be increased by solvent extraction vith 8-hydroxy-quinoline in 1 M CH_3COONa and $CHCl_3$ using an ammonium cetate buffer. Wavelength 309·3 nm. Reducing nitrous oxide-acetylene flame.

.10, 6.11, 6.12 CADMIUM, LEAD, MERCURY

A few colorimetric methods for Cd, Pb, and Hg have been published but inter-erences during the concentration steps are so likely that these methods are not to e relied on. The AAFS method with or without solvent extraction for cadmium nd lead is the most acceptable method at present, although interferences due to aboratory dust may be serious. Mercury may be determined without a flame. For cadmium good results are obtained with a graphite furnace. No details are iven here, but Table 6.2 may be of some use for setting up the routine.

Table 6.2. Characteristics for AAFS methods for Cd, Pb and Hg.

	Cadmium	Lead	Mercury
Detection limit			
direct aspiration	0·01 mg l^{-1}	0·05 mg l^{-1}	0·05 mg l^{-1}
solvent extraction	0·001 mg l^{-1}	0·001 mg l^{-1}	
Wavelength	228·8 nm	283·3 nm	253·7 nm
Fuel	Acetylene	Acetylene	special cell
Oxidant	Air	Air	
Type of flame	Oxidising	Oxidising	
References	Fishman and Erdmann, 1975		Goulden and Afghan, 1970

Organic mercury compounds are firstly oxidised to the inorganic mercury compounds by heating with H_2SO_4, $KMnO_4$ and $K_2S_2O_8$. The inorganic mercury is reduced with stannous sulphate in a hydroxylamine sulphate-sodium chloride solution to elemental mercury, which metal is sufficiently volatile to be flushed with a stream of air through a special cell put in place of the burner (no flame is needed)

CHAPTER 7

ORGANIC SUBSTANCES

7.1 ORGANIC CARBON

LEVEL III

Principle

The first stage in the analysis of organic carbon in fresh water is the oxidation of the carbon quantitatively to CO_2. The next stage is the transfer of the CO_2 produced, quantitatively, to an apparatus for CO_2 determination. The last stage is the measurement of CO_2.

The oxidation may be carried out by chromic acid in H_2SO_4 (§ 7.1.1), by persulphate (§ 7.1.2), by combustion in O_2 (§ 7.1.A) or by photochemical oxidation (§ 7.1.B).

Oxidation of organic compounds in water samples is generally complete, but for some artificial organic compounds the recovery by wet combustion methods may be incomplete. These compounds are, however, unlikely to occur in natural water samples. It is essential to check the technique by using a standard of oxalic acid, $(COOH)_2$. See § 7.3.1.

Interferences

Impurities in the apparatus may easily give high results; adsorption of CO_2 on the walls of the apparatus and by cleaning solutions may lead to low results. Chlorine, and acid oxides produced during the oxidation will interfere with CO_2 measurement and must therefore be removed. The dimensions of the equipment depend on the concentrations involved. If the dimensions are small CO_2 is removed more quickly than it would be from a larger apparatus.

Sampling and storage

Only specially cleaned glass-fibre filters are permissible for filtration. If these are to be used later for the determination of particulate carbon they should be preheated overnight at 500°C, with the pads spread out on an aluminium tray in the oven. They should be stored in a dust-free container. The sample should be filtered as soon as possible after sampling, to avoid decomposition of organic compounds.

There is no ideal method of storage. The addition of 2 ml H_2SO_4 (1 + 1) per 0·5 litre is the usual method and is effective for at least some weeks. Hydrolysis may occur, but at 5°C this process is slow. Deep-freezing is, in principle, the best method, but plastic containers may give off organic matter, and glass containers will sometimes break, so this method presents difficulties.

If no living plankton cells are present dissolved and particulate fractions may be separated by letting the sample settle for 24 hours in a glass-stoppered 500 ml bottle containing 2 ml H_2SO_4 (1 + 1). Aliquots of the supernatant liquid can then be removed, taking care not to disturb the sediment.

Dissolved organic carbon

7.1.1 Oxidation with chromic acid

Apparatus

Several apparatuses exist for the determination of dissolved organic carbon by chromic acid oxidation. These are too elaborate for a satisfactory description here, but excellent accounts of the apparatus and methods used are to be found in the original papers: Duursma (1961), Effenberger (1962) and Krey and Szekielda (1965). These methods are reliable but time consuming, and are useful to check results from other methods.

The apparatus used by Effenberger (1962) is shown in Fig. 7.1 to illustrate the principle of the method involved.

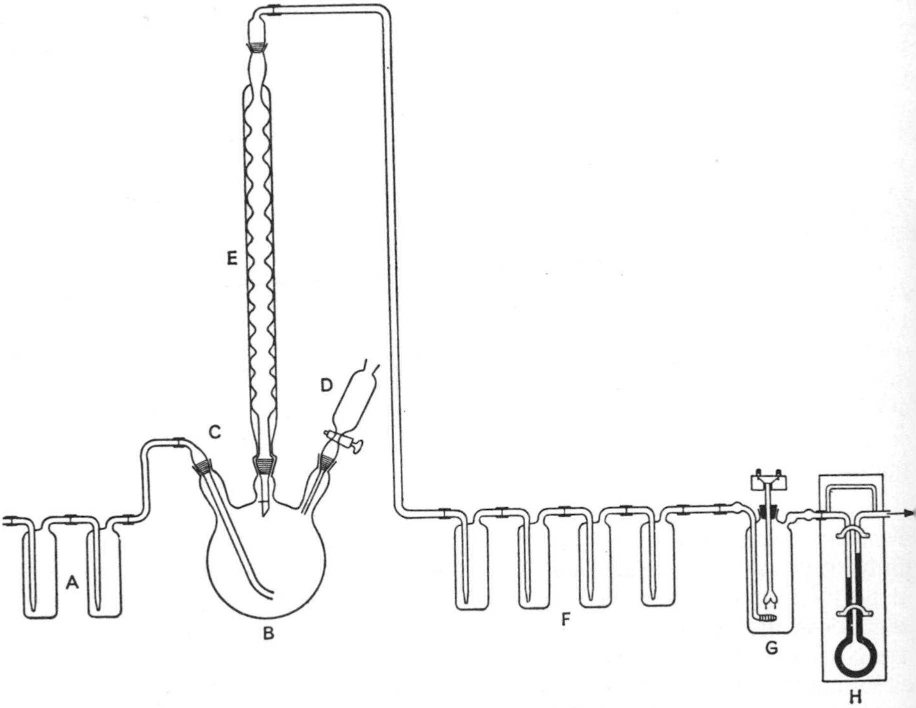

Figure 7.1. Apparatus for dissolved organic carbon determination according to Effenberger (1962).

A Wash-bottles for CO_2 absorption from the carrier gas (air). Contents: 50% KOH solution.
B Reaction vessel.
C Air-inlet.
D Oxidation mixture.
E Reflux-condenser.
F Wash-bottles, containing successively $K_2Cr_2O_7+H_2SO_4$, $KI+H_2SO_4$, $Na_2S_2O_3$, and concentrated H_2SO_4.
G Conductivity absorption cell.
H Manometer.

Reagents

A H_2SO_4 (1 + 1) See § 1.3.4.

B Oxidation mixture

Dissolve 10 g of $Ag_2Cr_2O_7$ and 42 g of $K_2Cr_2O_7$ in 400 ml of H_2SO_4 (A.R. s.g. 1·84) at about 90°C. Free the mixture of carbon by heating at 125°C for 8 hours while passing a stream of CO_2-free and H_2O-free O_2 through it at a constant rate. Avoid a higher temperature as auto-oxidation causes a precipitate in the mixture.

Fine crystals of $Ag_2Cr_2O_7$ can be obtained by adding slowly 50 g of $K_2Cr_2O_7$ in 800 ml of H_2O to a boiling solution of 50 g of $AgNO_3$ in 500 ml of H_2O (to which 4 ml conc. HNO_3 have been added). Collect the precipitate on a sintered glass filter and dry the precipitate at 100°C. (Commercial products must sometimes be purified by heating in H_2SO_4 at 125°C for 8 hours.)

Procedure

Place the sample in the reaction vessel (see Fig. 7.1) with 0·04 ml H_2SO_4 (A) per ml of sample. Drive off the inorganic CO_2 by heating for 1½ minutes just below boiling temperature. This CO_2, plus the O_2 introduced at a constant rate, should be allowed to escape at once from the reaction vessel. The O_2, freed from CO_2 by passing through sodium asbestos or sodium hydroxide and from H_2O by passage over $CaCl_2$, is passed into the apparatus at a constant speed. A suitable rate must be determined. Close the reaction vessel again immediately, so that no CO_2 can enter. No oxidation mixture should be introduced until the CO_2 absorption equipment shows that the apparatus is in equilibrium.

Now add to the sample an equal volume of the oxidation mixture. The temperature will rise to about 100°C. Then heat further to 130°C.

Be very careful that this procedure is always reproduced as exactly as possible on every occasion.

Keep all conditions constant until the CO_2 absorption equipment shows no more CO_2 uptake.

Range of application and accuracy

Dissolved organic carbon at concentrations from 0·2 to 20 mg l^{-1} of C can be determined with an accuracy of ± 0·03 mg l^{-1} of C, if complicated apparatus and techniques are used.

Note

Each series of determinations must start with blanks and standards to check that the apparatus is working correctly. A repeatable blank can be obtained only if details of operation are carefully standardised. For this reason, continuous working is preferable.

Liquids suck back easily if either the O_2 supply or the heating of the reaction vessel is irregular.

7.1.2 Oxidation with persulphate

This method is less time consuming than that of § 7.1.1.

Apparatus

The apparatus used by Menzel and Vaccaro (1964) is shown in Fig. 7.2.

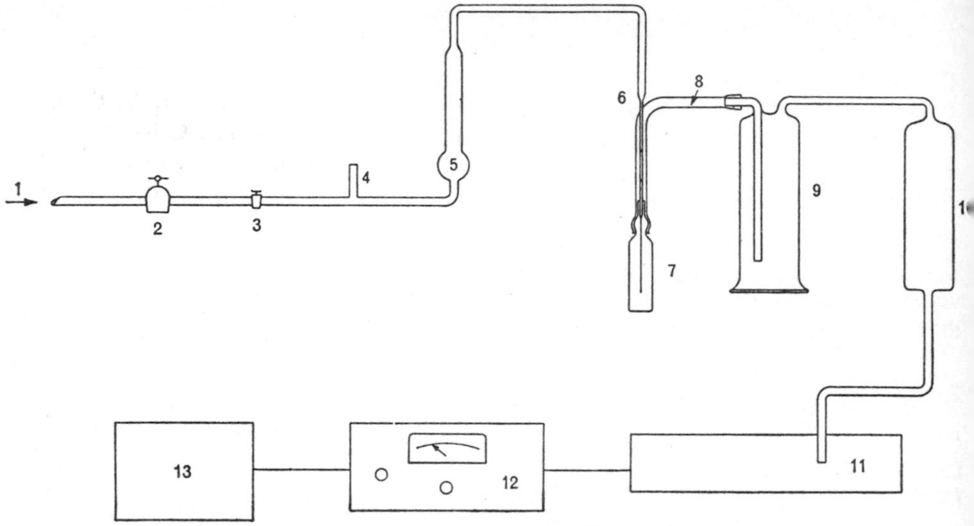

Figure 7.2. Schematic diagram of equipment used in the detection of CO_2, resulting from the oxidation of dissolved organic carbon, Menzel and Vacca.o (1964).

1. Nitrogen supply.
2. Regulator.
3. Needle valve.
4. Flow meter
5. Ascarite tube.
6. Hypodermic needle.
7. Ampoule.
8. Rubber tubing.
9. Gas washing bottle.
10. $Mg(ClO_4)_2$ drying tube.
11. Infra red CO_2 analyzer.
12. Amplifier.
13. Recorder.

The 10 ml pyrex ampoules are heated in a muffle furnace at 500°C for 4 hours to remove any traces of organic carbon. Alternatively the ampoules are heated with a solution of 40 ml of 3% H_3PO_4 and 20 g of $K_2S_2O_8$ at 130°C in an autoclave for two hours.

Reagents

A H_3PO_4, 3%
Dissolve 3 ml H_3PO_4 (A.R.) in 100 ml double distilled H_2O.

B $K_2S_2O_8$, dry

Procedure
Inject 5 ml of filtered sample into the 10 ml ampoule. Add sucessively 0·1 g $K_2S_2O_8$ and 0·2 ml 3% H_3PO_4. Bubble N_2 gas, free of CO_2, through a 10 cm cannula into the ampoule at a rate of 200 ml min^{-1} for 3 minutes to remove all inorganic carbon. Place a vaccine stopper, with a cut off fine hypodermic needle inserted through it, on top of the ampoule to restrict gas movement. Seal the ampoule in an oxygen-gas flame immediately. This operation is critical, as products of combustion from the flame may enter the ampoule. Details are given by Menzel and Vaccaro (1964).

Heat a batch of these samples at 130°C in an autoclave. About half an hour is normally sufficient, but for polluted waters, containing unusually stable compounds, a longer time may be necessary. The ampoules may at this stage be transported and the CO_2 subsequently transferred (§ 7.1.3) and estimated with a sensitive CO_2 detector e.g. the infrared CO_2 analyser (§ 7.1.7). Storage of the samples is no problem.

7.1.A Oxidation by combustion in O_2

Principle

The organic material is oxidised at a high temperature, after evaporation of the water sample. See Skopintsev (1960). This method has been applied more to the determination of particulate carbon (see § 7.1.8).

7.1.B Photochemical oxidation

Principle

Liquid samples are passed through a silica coil where they are irradiated with UV light in the presence of O_2. Organic carbon is oxidised to CO_2, which is measured by an infra-red gas analyser.

For further details see Baker *et al* (1974).

7.1.3 Transfer and purification of CO_2

The CO_2 formed during the oxidation must be transferred to the detector by flushing with O_2 or N_2.

During this transfer Cl_2, oxides of nitrogen and sulphur (formed during the oxidation § 7.1.1 only) and water-vapour must be removed.

This is achieved by passing the gaseous mixture from the reaction vessel successively through a KI solution (40 g in 50 ml 2 M H_2SO_4), metallic antimony and finally a silver-manganese catalyst. $Mg(ClO_4)_2$ or H_2SO_4 can be used to dry the CO_2.

The silver-manganese catalyst is prepared as follows:

Dissolve 40 g $MnSO_4.4H_2O$ in 200 ml double-distilled H_2O. Add a solution of 20 g $AgNO_3$ in 50 ml H_2O to which has been added 20 ml concentrated NH_4OH. Filter through a sintered glass filter, wash the precipitate free of SO_4^{2-} and dry at 120°C. Decompose at 400–500°C (avoid heating at greater than 550°C) for 2 hours.

All reagents should be of high purity. Regular replacement is necessary. Solutions must be made with double or triple distilled H_2O (distillation from H_3PO_4 + $KMnO_4$ solution). Those ground glass joint surfaces which are in contact with the oxidation mixture or H_2SO_4 should be sealed with H_2SO_4 (s.g. 1·84)

Special procedure for CO_2 transfer following persulphate combustion

The neck of the sealed ampoule is inserted into the rubber tubing (Fig. 7.2) and N_2 is flushed through the system until the IR analyser indicates the complete removal of extraneous CO_2. The gas flow is then shut off, the tip of the ampoule is crushed, and the cannula is inserted into the solution. The gas flow is started again and kept constant. The gas is washed by the acid KI solution and dried by $Mg(ClO_4)_2$.

7.1.4 Determination of CO_2 by absorption in excess $Ba(OH)_2$; back-titration with HCl (Kay, 1954).

Reagents

A ½ $Ba(OH)_2$, 0·05 M
Dissolve 16 g of $Ba(OH)_2.8H_2O$ and 20 g of $BaCl_2$ in 2 litres of boiled H_2O. Store the solution in a container which is connected directly to a burette. Exclude CO_2 from both with soda lime tubes. Allow any $BaCO_3$ precipitate to settle before transferring the solution to the burette.

B HCl, 0·025 M
Dilute 0·100 M HCl (§ 1.3.4) (250 ml → 1000 ml).

C Mixed indicator
Mix 10 ml of 0·1 % Thymol Blue (in 50 % alcohol) with 30 ml of 0·1 % phenolphthalein in 50 % alcohol. At pH = 9 the colour changes from red-violet (alkaline) to clear yellow (acid).

Procedure
Backtitrate the excess of $Ba(OH)_2$ with HCl using the mixed indicator to locate the end point.

Note
The $Ba(OH)_2$–HCl titration method can only be used for waters which have a relatively high concentration of organic carbon. 1 ml of HCl is equivalent to 0·15 mg of C per sample. No direct check for completion of the reaction is possible.

7.1.5 Determination of CO_2 by absorption in excess NaOH; change in conductance (Effenberger 1962; Szekielda and Krey 1965)

The CO_2 is absorbed in a solution of NaOH or $Ba(OH)_2$. This causes a decrease in the conductance of the solution. NaOH is preferable because there is then no precipitate on the electrodes, or on the container surfaces, although the change in conductance is smaller than with $Ba(OH)_2$. The concentration of NaOH or $Ba(OH)_2$ should be between 0·01 M and 0·02 M.

The two fixed platinum electrodes in the absorption vessel (see Fig. 7.1) form one arm of a Wheatstone bridge. To measure the conductance an A.C. voltage of approximately 10 volts can be used. The frequency should be not less than 1000 cycles per second. The temperature must be kept constant.

The apparatus must be standardised empirically at the same temperature. A standard of $(COOH)_2$ made with boiled H_2O (see § 7.3.1) can be used. There should be a linear relationship between CO_2 and conductance.

7.1.6 Determination of CO_2 by coulometric titration (Duursma, 1961)

As with the conductometric determination, this method makes it possible to follow the arrival of CO_2 from the combustion chamber continuously. The advantage of the coulometric method is that conditions in the apparatus are always constant, and the apparatus is immediately ready for the next analysis.

The CO_2 is absorbed in a $BaCl_2$ solution which is maintained at a fixed pH of 9·6 by electrolysis. The anode is separated from the absorption solution by an agar salt bridge.

By measuring the amount of electricity (current multiplied by time), the amount of absorbed CO_2 can be calculated. Absorption of 1 mg carbon as CO_2 requires 16·084 coulomb (i.e. 804·2 seconds of electrolysis with a constant current of 20 mA), to maintain the pH constant.

By plotting quantity of electricity against time the course of CO_2 absorption can be followed.

7.1.7 Determination of CO_2 by infrared gas analysis (Montgomery and Thom, 1962; Menzel and Vaccaro, 1964)

If only small amounts of CO_2 are available for determination, a particularly sensitive technique is necessary. Measurement of absorption of infra-red radiation by CO_2 in the gas mixture is a suitable technique. Following the persulphate oxidation (§ 7.1.2) no other method of CO_2 estimation is sufficiently sensitive, but it can of course also be used following chromic acid oxidation (§ 7.1.1).

The measurement can be made either directly in the purified gas stream or after collecting the CO_2 in an evacuated container. The IR analyser must be calibrated empirically (Montgomery and Thom, 1962).

The use of a commercially available apparatus is described by Busch (1967).

7.1.8 The determination of particulate carbon by combustion in O_2. (See Szekielda and Krey, 1965).

Principle

The particulate material is collected on a glass-fibre filter. After drying, the filter is heated at a high temperature in a stream of O_2, and the liberated CO_2 is determined conductometrically. (The method can also be used in combination with an infrared gas analyser). The CO_2 should then be absorbed as described in § 7.1.7. Oxygen is used instead of nitrogen as the carrier gas (Menzel and Vaccaro, 1964).

Accuracy, sensitivity and range of application

The range of application is 50–1000 $\mu g\ l^{-1}$ C. The standard deviation of 10 samples containing 0·34 mg l^{-1} C was 15 $\mu g\ l^{-1}$ C. Accuracy is about 4%.

7.1.9 Automatic apparatus

There are many commercial instruments for the measurement of carbon and some for COD. Details are given in Department of the Environment, Notes on Water Pollution no. 59, 1972. The principle of operation of a few more recent

Table 7.1. Some recent instruments for the analysis of carbon or COD in water.

Name	Principle	Measures
CMA	Wet persulphate digestion in ampoules	C
Delta Scientific	High temperature oxidation followed by nephelometry	C
Dohrmann	High temperature oxidation, then measurement by flame ionisation detector	C
Carlo Erba	High temperature oxidation, then measurement by flame ionisation detector	C
Ionics	High temperature oxidation, then measurement by flame ionisation detector	C
Axel Johnson	Wet oxidation by chromate, then photometric measurement of Cr^{3+}	COD
Maihak	UV photolytic oxidation, then infra-red gas analysis.	C

instruments is given in Table 7.1. The instruments commonly measure total carbon, dissolved organic carbon and inorganic carbon, but not all instruments measure all these.

7.2 BIOCHEMICAL OXYGEN DEMAND (BOD)

ALL LEVELS

Principle

The rate of removal of O_2 by organisms using the dissolved or even particulate organic matter in water may be determined and is often thought to be a measure of the concentration of organic matter present. This measurement, called biochemical oxygen demand or BOD, is especially useful as an easy, but approximate, index of organic pollution (or of sewage treatment plant efficiency). The oxygen consumed during algal respiration is included and, in some cases, this may account for up to 50% of the total BOD.

The BOD test is commonly made by measuring O_2 concentration in samples before and after incubation in the dark at 20°C for 5 days. Preliminary dilution and aeration of the sample are usually necessary to ensure that not all the O_2 is used during the incubation. Excess dissolved oxygen must be present during the whole incubation. Samples absorbing more than 6 mg l^{-1} of O_2 should therefore be diluted with a synthetic dilution water made from BOD-free water to which the major constituents are added in the same concentration as in the samples. Sometimes a culture of bacteria is added so that more of the organic matter will be used up during the incubation.

(For detailed instructions see Standard Methods (1965).)

The importance of short experimental times and the stoicheiometry of the reactions involved is stressed by Busch (1966).

In waters that are not heavily polluted, the bacterial degradation of the organic matter is slow and often continues for 10, 20, or more days (Straškrabová-Prokešová, 1966). The standard BOD method is not usable under these conditions, but a simplified method can give some information about differences between lakes or rivers or about seasonal changes in the dissolved organic matter in one water body (Stangenberg, 1959).

Procedure

Adjust the sample pH to the range 6·5 to 8·5.

If the O_2 content of the original sample is very low, the sample should be aerated for 5 or 10 minutes. Place portions of the sample into 3 glass stoppered bottles of either 125, 250, or 300 ml capacity and immediately determine the O_2 concentration in one bottle (see § 8.1.1 or § 8.1.2). Incubate the remaining 2 bottles in the dark at a standard temperature (e.g. 20°C) or at the temperature of the original sample for 1 to 5 days. Then determine the O_2 concentration in the bottles. Subtract the mean value from the original value of O_2 to give BOD.

A considerable part of the BOD value may result from oxidation of ammonia. Nitrifying bacteria are particularly sensitive to several trace elements, so nitrification in polluted waters may be erratic and not reproducible. Nitrification can be inhibited by adding 1 ml of a solution of 0·5 g l^{-1} of allylthiourea.

Notes

(1) Samples should be free from preservatives. Vigorous shaking must be avoided. Mixing with dilution water must be done by repeated inversion of bottles, avoiding the formation of air bubbles.

(2) A dissolved oxygen probe, suitable for BOD measurements and using a standard pH meter, is made by EIL.

7.3 CHEMICAL OXYGEN DEMAND (COD)

LEVEL II

COD DUE TO DISSOLVED ORGANIC COMPOUNDS

Principle

The concentration of organic compounds in lake water can be estimated by their oxidability by substances such as $K_2Cr_2O_7$ and $KMnO_4$.

Organic carbon and O_2 are indirectly related because the reaction is:

$$C_xH_{2y}O_z + \left[x + \frac{(y - z)}{2}\right] O_2 \rightarrow x\,CO_2 + y\,H_2O$$

The amount of carbon oxidized can therefore be calculated from the amount of O_2 taken up only if the ratio of $(y - z)$ to x is known. This ratio is, however, in most cases unknown. The COD method is, however, simpler than a carbon determination, and is directly comparable to BOD which estimates the organic compounds available for bacterial activities. COD also has particular significance when photosynthesis is expressed in the same units; as mg l^{-1} O_2 (Golterman, 1967a).

$K_2Cr_2O_7$ is the most suitable oxidising agent. The reaction is:

$$Cr_2O_7^{2-} + 14H^+ + 6e^- \rightarrow 2Cr^{3+} + 7H_2O$$

so that stoicheiometry requires $\frac{1}{6}$ $Cr_2O_7^{2-}$.

It cannot, however, be used for samples containing more than 2 g l^{-1} of Cl^- because of the oxidation of Cl^- to Cl_2. In these cases $KMnO_4$ must be used, though the results are more variable because $KMnO_4$ is self oxidizing. When $K_2Cr_2O_7$ is used $HgSO_4$ must be added to mask the chloride in the range 0·2 to 2 g l^{-1} of Cl^-.

Ag_2SO_4 is used as a catalyst for the oxidation. It is particularly effective for short straight-chain alcohols and acids.

The oxidation is carried out with an excess of $K_2Cr_2O_7$. The COD is estimated either by the determination of unused $Cr_2O_7^{2-}$ (§ 7.3.1) or by the spectrophotometric determination of the Cr^{3+} formed (§ 7.3.2).

(See Burns and Marshall, 1965, Maciolek, 1962.)

Sampling and storage

The water samples should be taken with water-sampling bottles which do not release organic substances into the water. Filtration through glass-fibre filters is recommended, but hard paper filters may be used if the sample has a high COD. The filters should be prerinsed with water. The glass-fibre filters should be pretreated as described in § 7.1. (Particulate COD may be measured; see § 7.3.4.)

The analysis should be made as soon as possible and the samples kept cold (4–8°C) beforehand. For long storage deep-freezing is recommended.

Apparatus

Autoclave and 250 ml flat bottom flasks with covers, or 250 ml flat bottom flasks with ground-glass connections to 30 cm Liebig (straight tube single surface) condensers.

For method 1: Dead stop titrator (see § 9.13) or pH meter.

For method 2: Spectrophotometer or colorimeter with filters with maximum transmittance close to 500 nm and 585 nm and with narrow band with a sharp cut off.

7.3.1 Volumetric

Oxidability by $K_2Cr_2O_7$.

Reagents

A Standard oxalic acid, $\frac{1}{2}$ $(COOH)_2$ 0·0125 M (10 ml = 1 mg COD)

Dilute 0·100 M $\frac{1}{2}$ $(COOH)_2$ (see § 1.3.4) (25 ml → 200 ml).

B $\frac{1}{6}$ $K_2Cr_2O_7$, 1·000 M

Dissolve 49·035 g of $K_2Cr_2O_7$ (A.R., dried for 2 hours at 105°C) in H_2O (double distilled) and dilute to 1000 ml.

C_1 H_2SO_4 (s.g. 1·84, A.R.)

C_2 H_2SO_4 (1 + 1) A.R. See § 1.3.4

D Ferrous solution, 0·25 M.

Either D_1 or D_2 may be used.

D_1 Fe–ES solution, 0·25 M

Dissolve 95·5 g of Fe-ethylenediamine sulphate $4H_2O$ (available from Merck or Baker) in H_2O, add 1 ml of H_2SO_4 (1 + 1) and dilute with H_2O to 1000 ml.

D_2 $Fe(NH_4)_2(SO_4)_2$, 0·25 M

Dissolve 98 g of $Fe(NH_4)_2(SO_4)_2.6H_2O$ in H_2O. Add 40 ml of H_2SO_4 (1 + 1) and dilute to 1000 ml.

Standardise the Fe solution (D_1 or D_2) against $K_2Cr_2O_7$ solution (B).

E NH_2SO_3H (Sulphamic acid), (dry, A.R.)

F Diphenylamine indicator, 0·5%

Dissolve 0·50 g of diphenylamine in 20 ml of H_2O and add, with caution, 100 ml of H_2SO_4 (s.g. 1·84).

G Ag_2SO_4 (dry, A.R.)

H $HgSO_4$ (dry, A.R.)

I Oxidation mixture

Add 200 ml H_2SO_4 (C_1), cautiously with stirring, to a solution of 1 g Ag_2SO_4 (G) in 100 ml $K_2Cr_2O_7$ (B).

Procedure

Mix 70 ml of sample containing not more than 16 mg COD (as O_2) or an aliquot diluted to 70 ml, in a 250 ml flask with powdered $HgSO_4$ (H) in the ratio $HgSO_4$: $Cl^- = 10:1$ (Cl^- less than 2 g l^{-1}). Add, with caution, 30 ml oxidation mixture (I). Attach the condenser to the Erlenmeyer flask, and put the flask in a water bath. Either boil for 3–6 hours or autoclave the covered flasks at about 135°C (3 atmospheres pressure) for 3–6 hours. Cool the flask, and then:

EITHER

Titrate the excess $K_2Cr_2O_7$ with reagent (D_1) using a dead stop end point titrimeter or a potentiometric titrimeter (see § 9.12 and § 9.13). Graph the meter readings against volume of reagent.

OR

Add 1 ml of indicator (F) and titrate the excess $K_2Cr_2O_7$ with reagent (D_2). At the end point the colour changes sharply from turbid blue to brilliant green.

Run a blank using 70 ml of double-distilled H_2O, and all reagents.

Calculation

Let x = COD of the water sample [mg l^{-1}]

then $$x = \frac{8{\cdot}0\, C_t\,(V_{ab} - V_{as})}{V_s}$$

where C_t = concentration of titrant [mmol l^{-1}]
V_{ab} = volume of titration of the blank [ml]
V_{as} = volume of titration of the sample [ml]
V_s = volume of water sample [ml]

Precision and accuracy

It will be apparent that extreme care in the titration is necessary. The titration error is about $\frac{1}{4}$ mg COD per sample or 4 mg l^{-1} COD. The precision depends primarily on the difference $V_{ab} - V_{as}$ between two relatively large volumes, and if this difference is small (= low COD) the measurement is imprecise. The use of a piston microburette makes the titration more precise. The capacity of a piston microburette is usually not more than 5 ml, so the solution must in this case be made up to volume after the oxidation, and an aliquot taken for titration.

7.3.2 Colorimetric

Oxidability by $K_2Cr_2O_7$ (measurement of Cr^{3+} produced: de Graaf and Golterman, 1967).

Reagents

See § 7.3.1 but without (D) and (F).

Procedure

Mix the sample, $HgSO_4$, and an equal volume of the single oxidising solution (I) in a 100 ml volumetric flask. Allow to cool and make up to the mark with H_2SO_4 (C). Depending on the sample volume use C_1 or C_2. Mix. Transfer the solution to a 250 ml flask, cover the flask with an inverted beaker and autoclave for 2 hours at

120–130°C. Alternatively, boil the solution gently on a hot plate until the volume is reduced by 50% (about 2 hours). The temperature should not rise above 124°C (see back cover). Cool and make up to volume. Run a blank as in procedure § 7.3.1.

When cool, but within 3 hours measure the absorbance of reagent blanks, standards and samples against a H_2O blank in suitable cells at wavelengths as close as possible to both 500 nm and 585 nm.

If the solutions are turbid they should be filtered before the absorbance is measured.

Calibration curve

Prepare a calibration curve covering the range 1 up to 16 mg COD by using known amounts of $(COOH)_2$ (A) and using exactly the same method as described above. The calibration curve should be identical with a calibration curve using known amounts of a standard solution of $K_2SO_4.Cr_2(SO_4)_3$ boiled for the same time in H_2SO_4 under the same conditions.

Calculation

The absorbance of the sample at 585 nm has two components. The major one is due to the Cr^{3+} produced in the reaction. The minor one is due to the unused $K_2Cr_2O_7$, the concentration of which is lower in the sample than in the blank. The decrease in absorbance due to this cause is determined at 500 nm because Cr_3^+ does not absorb at this wavelength. If a colorimeter is used a check should be made to ensure that the absorbance of Cr^{3+} at 500 nm is 0. (If not try a filter with maximum at 490 nm.)

Let x = absorbance of Cr^{3+} produced

then $$x = A_{585} - \frac{B_{585}\,A_{500}}{B_{500}}$$

where A = absorbance of sample
B = absorbance of blank

and subscripts 500, 585 indicate measurements at 500 nm and 585 nm.

Precision

The absorbance produced by a sample containing 1 mmol l^{-1} COD in a 5 cm cell is about 0·030. The precision is about 1 mg l^{-1} of O_2 (corresponding to $x = 0{\cdot}005$) which can be improved by a factor of 4 when using a 20 cm cell.

Note

The colour of the Cr^{3+} ion is stable for 3 hours; afterwards the violet hydrate may be formed in significant amounts.

Interferences: method § 7.3.1 and § 7.3.2.

Fe^{2+} and H_2S interfere and must be removed by bubbling air through the sample.

NO_2^- interferes, but can be removed by adding 10 mg of sulphamic acid (E), per mg of NO_2–N to the dichromate solution.

7.3.3 Volumetric

Oxidability by $KMnO_4$.

For brackish waters the alkaline oxidation method is often applied, in spite of difficulties with incomplete oxidation and reproducibility. Cl^- interferes less than in the acid-dichromate method.

The method should always be used by analysing batches of a number of samples together with blanks and standards.

Apparatus

Water bath, with lid, containing COD-free water, capable of taking, say, 20 Erlenmeyer flasks on a rack. Ordinary tap water preboiled with $KMnO_4$ should be COD free.

Reagents

A Standard oxalic acid, $\frac{1}{2}$ $(COOH)_2$, 0·100 M See § 1.3.4.

B $\frac{1}{5}KMnO_4$, 0·1 M

Dissolve 3·16 g of $KMnO_4$ + 16 g of NaOH in double or triple distilled H_2O. Dilute to 1 litre. Allow to stand in the dark for several days. Filter through a glass filter and standardize against $(COOH)_2$. Store in a refrigerator.

C $Na_2S_2O_3$, 0·1 M

Dissolve 15·81 g of $Na_2S_2O_3$ in 1 litre of H_2O. Add 1 g of Na_2CO_3 as preservative. Allow the solution to stand for a day before standardizing.

D H_2SO_4, 2 M See § 1.3.4

E KI solution, 10%

Dissolve 10 g of KI in 90 ml of H_2O.

F Starch solution, 1%

Heat 1 g of soluble starch in 100 ml H_2O until it dissolves.

Procedure

Add 50 ml of sample or an aliquot containing not more than 0·8 mg COD (16 mg per litre), to a 100 ml Erlenmeyer flask. With each series run two or three blanks (double distilled H_2O) and at least two standards. Heat the water bath to boiling point. Add 5·0 ml of $KMnO_4$ (B) to each flask and place the batch in the water bath for exactly one hour. Cool for 10 minutes. Add 5 ml of KI (E) and directly afterwards 10 ml of 2 M H_2SO_4 (D). Titrate with thiosulphate (C), and when the solution has become pale yellow add 1 ml of the starch solution. Continue the titration until the blue colour disappears.

If the sample titration volume is more than 90% of the blank titration volume then the determination should be repeated with $KMnO_4$ and $Na_2S_2O_3$ solutions diluted to $\frac{1}{2}$ or $\frac{1}{10}$ the original value.

Calculation

As for method § 7.3.1.

CHEMICAL OXYGEN DEMAND DUE TO PARTICULATE MATTER

LEVEL II

7.3.4 Oxidability

If more than about 25–50% of the total organic matter is present in particulate form it is possible to determine the COD in the total sample and in the filtrate (see § 7.3.1 or § 7.3.2). The difference gives the COD of the particulate matter. If the particulate matter is present in smaller proportion, or if the highest accuracy must be obtained, the particulate matter should be collected on a glass-fibre filter, and the filter can then be introduced in a quantity of double distilled water (Golterman, 1971) for analysis by method § 7.3.1 or § 7.3.2. The filter should be pretreated as described in § 7.1.

7.3.5 Method deleted from this edition

LEVEL III

7.3.6 Automatic methods

The methods § 7.3.1 or § 7.3.2 can also be used at level III, using a recording potentiometric device, and can be completely automated. Some instruments are commercially available for the determination of COD. See § 7.1.9 and Table 7.1.

7.4 DISSOLVED ORGANIC COMPOUNDS

LEVEL III

By fluorescence in ultraviolet light

Principle

The method of determination of the fluorescence of water is useful for characterising water bodies during mixing processes. This fluorescence is a qualitative indicator for the presence of dissolved organic matter with fluorescent groups which are relatively resistant to decomposition. The relationship between the strength of fluorescence and the concentration and composition of organic matter is, in most instances, poorly known.

Procedure

Irradiate the filtered water sample with UV light in a quartz cell. The fluorescent light may be measured by a photocell or photomultiplier. There are several instruments that give satisfactory results, for example those made by

Carl Zeiss	Oberkochen, W. Germany.
Baird-Atomic	Bedford, Mass., USA.
Perkin-Elmer	Norwalk, Conn., USA.
Turner Ass.	Palo Alto, Calif., USA.
Aminco	Bethesda, Maryland, USA.

Calculation
The fluorescence measured is only proportional to the concentration of fluorescent substances at high dilution. In order to express the fluorescence of natural waters in comparable numbers the empirical mFl unit of Kalle (1955/1956) is recommended: a solution of 1 mg of quininebisulphate in 1 litre 0·005 M H_2SO_4 gives a fluorescence of 700 mFl. A correction should be applied (Duursma and Rommets, 1961; Duursma, 1961, 1974).

7.5 CARBOHYDRATES

LEVEL II

7.5.1 Colorimetric

Principle
Carbohydrates are estimated using the colour produced in the reaction of the furan derivatives of the sugars, formed by dehydration with strong H_2SO_4 at 100°C, with anthrone in strong H_2SO_4. Prior hydrolysis is not necessary as this takes place during the development of the colour. The final colour is dependant on time and temperature.

Different carbohydrates react at different rates with the reagent (Yemm and Willis, 1954). For example, after five minutes pentoses yield an absorbance of about 50%, and fructose 200%, of the absorbance of glucose, the usual standard. Although glucose reaches a usable absorbance after 5 minutes heating, 15 minutes is chosen as the recommended heating time because the absorbances of the different sugars differ less after this period. Solutions must be cooled rapidly after colour development.

Thiourea is added to improve the stability of the anthrone. If there is doubt that glucose is the main sugar, use the method of § 7.5.3.

Reagents

A Glucose standard solution, 100 μg ml^{-1}
Dissolve 100·0 mg of glucose in 1000 ml of H_2O. This solution will keep for short periods, but only if stored at 0°C. Just before use prepare dilutions containing 10, 25, 50 and 100 μg ml^{-1} of glucose.

B Anthrone solution, 0·1%
Add 500 ml of H_2SO_4 (s.g. 1·84) carefully, with stirring and cooling, to 200 ml H_2O. Cool to room temperature. Add 0·1 g anthrone and 1 g thiourea with stirring. Leave for 1 hour before use. The solution can be kept for up to 2 weeks in a dark bottle, in a refrigerator. Otherwise prepare a fresh solution daily.

Procedure (total sugar)
In a tall tube (18 × 180mm) layer 5 ml of a sugar solution, or of a suspension or extract of organisms, containing not more than 100 μg of glucose equivalent per ml, on 25 ml of cold anthrone reagent (B) directly from the refrigerator. Stopper the tubes loosely. Mix, and place the tubes together in a boiling water bath for exactly 15 minutes. Cool the tubes together in a beaker with running water as quickly as possible, or immerse them in an ice bath. Measure the absorbance of

the reagent blank, standards, and samples, against H_2O, in suitable cells within 3 hours using a wavelength between 620 and 630 nm.

Notes

Humic compounds react in the same way as glucose (Stabel, 1977).

The use of an ethanol containing anthrone solution has been described by Fales (1951) and Hewitt (1958). The ethanol is incorporated to stabilize the coloured products, but less is known about the reaction rates of the different sugars. The reagent is prepared as follows:

Dissolve 0·4 g of anthrone in 200 ml of H_2SO_4 (s.g. 1·84). Add this solution to a flask containing 60 ml of H_2O and 15 ml of 95% ethanol.

Acid soluble carbohydrates

Hydrolyse the dry cells, or a cell suspension, with a final H_2SO_4 concentration of 0·25 M in a boiling water bath for 30 minutes. Centrifuge (3000 rpm) and use 5 ml as described for the procedure above.

LEVEL II/III

7.5.2 Volumetric

Principle

Sugars and polysaccharides are hydrolysed to monosaccharides in HCl (Blažka, 1966b). The reducing capacity of the sugars is then determined by a micro ferricyanide back-titration (Hanes, 1929; Hulme and Narain, 1931). The ferrocyanide which is formed is precipitated with Zn^{2+}. The residual ferricyanide is estimated by adding excess KI (G) and acidifying. The ferricyanide is quantitatively reduced by KI and the I_2 formed titrated with $Na_2S_2O_3$ (E). The method estimates total reducing capacity and is not specific for carbohydrates.

The advantage of the ferricyanide method over the anthrone method is that the total reducing value of a mixture of sugars is approximately the sum of the different components. Comparison with procedure § 7.5.1 is useful.

Reagents

A Glucose standard solution, 100 μg ml^{-1}

See § 7.5.1, reagent (A).

B HCl, 1 M

Dilute 4 M HCl See § 1.3.4.

C CH_3COOH, 5%

Use a grade free of reducing substances.

Dilute 5 ml glacial acetic acid to 100 ml with distilled water.

D NaOH, 2 M

Dilute 4 M NaOH See § 1.3.4

E Thiosulphate solution, 0·01 M

Dissolve 2·5 g of $Na_2S_2O_3.5H_2O$ in H_2O. Add a pellet of NaOH or 1 g of Na_2CO_3 and dilute to 1 litre. Store in a brown bottle. Standardise (see § 8.1.2) and dilute to exactly 0·01 M.

F Alkaline $K_3Fe(CN)_6$ solution, about 0·025 M
Dissolve 8·25 g of $K_3Fe(CN)_6$ and 10·6 g of Na_2CO_3 in 1 litre of H_2O. Store in dark.

G KI solution, 2·5%
Dissolve 12·5 g of KI, 25·0 g of $ZnSO_4$, and 125 g of NaCl in H_2O and dilute to 500 ml.

H Soluble starch See § 8.1.2

Procedure
Heat the dry residues of water samples, or the collected organisms or precipitates, with 10·0 ml 1 M HCl (B) at 100°C for 30 minutes. The acid soluble carbohydrates are thus extracted, and sucrose is converted to glucose plus fructose. Add 5 ml 2 M NaOH (D), centrifuge or filter, mix 10·0 ml of extract with 5·0 ml ferricyanide solution (F) in a test tube. Cover the tubes and place them in a boiling water bath for 15 minutes. Cool for 3 minutes in cold running water and add 5 ml KI solution (G). I_2 is liberated and insoluble Zn ferrocyanide is formed. Add 3 ml acetic acid solution (C). Titrate the I_2 with $Na_2S_2O_3$ (E), as described in § 8.1.2. Run blanks through the whole procedure. Standards must be of volume 10 ml. Dilute standard (A) as necessary.

Calibration and calculation
Calculate the volume of thiosulphate used, V_t [ml,] from:

$$V_t = V_b - V_s$$

where V_b = volume of 0·01 M $Na_2S_2O_3$ used in blank titration [ml]
V_s = volume of 0·01 M $Na_2S_2O_3$ used in sample or standard [ml]

The calibration of V_t against mass of glucose in the range 0·1 to 3·0 mg should be linear, and because the reaction is nearly stoicheiometric the slope should be about 0·340 ml mg^{-1}, but the extrapolated intercept at zero mass is usually about 0·05 ml. Other sugars gave slopes of 0·341 (fructose), 0·338 (invert sugar), and 0·455 (maltose).

The mass of sugar in the sample, M_s [mg], is given by:

$$M_s = 1{\cdot}5\ M_a$$

where M_a is the mass [mg] in the 10 ml aliquot used.

Precision and accuracy
Accuracy of 0·01 mg is claimed for solutions of single sugars. The accuracy for unknown complex mixtures must obviously be less than that.

7.5.3 Colorimetric

Principle
Phenol and sulphuric acid are added directly to water and the resultant absorbance read at 490 nm (Strickland and Parsons, 1968). This method is very suitable for total carbohydrate analysis because the same equivalent colour is produced with different sugars.

Reagents

A Glucose standard solution, 100 μg ml^{-1}
See § 7.5.1, reagent (A).

B Phenol reagent, 5%

Distill phenol twice using all-glass apparatus.

Dissolve 25 g redistilled phenol in 500 ml H_2O. Store in an all-glass bottle in a refrigerator.

C Sulphuric acid reagent

Dissolve 2·5 g hydrazine sulphate in 500 ml concentrated H_2SO_4 (s.g. 1·84).

Apparatus

125 ml pyrex Erlenmeyer flasks.

Rapid-flow 50 ml burettes.

Procedure

Pipette 2 ml filtered water into a 125 ml Erlenmeyer flask and add 2 ml phenol reagent (B) from a burette. Add 10 ml H_2SO_4 reagent (C) from a rapid-flow burette, with mixing. (The working area should be well-ventilated). Cover the flask with foil and let it cool for 1 hour at room temperature. Measure the absorbance at 490 nm using a 10 cm glass cell.

Make a calibration curve for the range 5 to 40 μg glucose. Use reagent blanks.

Precision

Precision may be about 1%.

Lower limit of detection is about 0·4 μg glucose.

7.6 LIPIDS OR FATS

LEVEL II

Principle

The lipids are extracted from the particulate matter by a nonpolar or semipolar solvent. Lipids often contain compounds with double bonds which are easily oxidised, especially during drying. It is therefore better to use fresh material.

After evaporation of the solvent the lipids are determined gravimetrically or by a carbon determination (see § 7.1 or § 7.3).

The method is not specific. A specific determination would involve hydrolysis with KOH, methylation of the fatty acids produced, followed by a gas chromatographic determination. (See for example Farkas and Herodek, 1964; Otsuka and Morimura, 1966.)

Thin-layer chromatography is suitable for the separation of lipids. See Randerath, 1966, and Stahl, 1970.

7.6.1 Extraction with methanol and ethanol-ether

Reagents

A Methanol, reagent grade

B Ether, reagent grade, peroxide-free

Check for peroxides by shaking the ether with an equal volume of 10% KI solution. Leave to stand overnight. A yellow colour indicates the presence of peroxides. They can be removed by shaking the ether with an acid solution of $FeSO_4$. After discarding the water layer the ether is dried over Na_2SO_4 and then redistilled (but NOT OVER A NAKED FLAME AND NOT TO DRYNESS).

C Ethanol-ether, 1 + 1
Mix equal volumes ethanol (96%) and ether, both reagent grade but free from peroxides.

D Na_2SO_4, anhydrous

Procedure
Obtain the fresh particulate material by centrifuging. Homogenise larger animals in methanol. Add 5 ml methanol (A) and leave to stand for 20 minutes. Decant the methanol into a crystallising or evaporating basin and wash the particulate material successively with 5 ml of methanol (A) and twice with 5 ml of ethanol-ether (C). Combine and evaporate the extracts. Carefully extract the dry residue thus obtained with dry ether (B). Shake the ether with H_2O in a separatory funnel. Discard the water layer and dry the ether layer with Na_2SO_4. Filter the ether through a G3 sintered glass filter (15–40 μm pore size) into a crystallising dish, and evaporate under vacuum and at room temperature.

If sufficient lipids are present (about 5 mg) the final determination may be made gravimetrically. More sensitive is an oxidation with dichromate, preferably procedure § 7.1.1 + § 7.1.4 or § 7.3.2.

Note
All evaporations must be made in a vacuum desiccator or under N_2 in order to avoid oxidation of unsaturated fatty acids. If the *ether* extracts are being evaporated the vacuum must be applied slowly. Do not use plastic containers. Blanks are essential to ensure that solvent has been completely removed.

7.6.2. Colorimetric

Principle

Vanillin reacts with the breakdown products of lipids to give a pink colour (Chabrol and Charonnat, 1937; Chabrol, Böszörményi and Fallop, 1949). To avoid oxidation of the double bonds of the fatty acids during the evaporation stage the procedures described by Zöllner and Kirsch (1962), Mittelholzer (1970), Christie (1973) and Vijverberg and Frank (1976) are used. A standard closely resembling the composition of fatty acids in zooplankton (Farkas and Herodek, 1964) is used.

Reagents

A H_2SO_4 (s.g. 1·84, A.R.)

B Na_2SO_3, about 5%
Dissolve 5 g of Na_2SO_3 in 100 ml of H_2O. Prepare freshly.

C Phosphovanillin reagent
Dissolve 0·6 g of vanillin in 100 ml of H_2O. Mix with 400 ml of H_3PO_4 (s.g. 1·75, A.R.).

D Methanol-chloroform (1:2)
Mix 0·5 litre of methanol (laboratory grade) with 1·0 litre of chloroform (laboratory grade). Store in a dark bottle.

E Standard fatty acid solution

Dissolve 8·0 mg of palmitic acid and 24·0 mg of linoleic acid in 100 ml of solvent (D). Dilute 50 times for daily use (5 ml → 250 ml).

Procedure

Extract the lipids by homogenizing a zooplankton sample with 10 or 15 ml of methanol-chloroform mixture (D). Filter the suspension through a G4 glass filter and transfer suitable volumes of the filtrate to test tubes in a vacuum desiccator. Shade from direct light and place in a thermostatically controlled waterbath at 55°C. Evaporate the solvent under low pressure (about 5 cm Hg), while flushing with N_2 washed through solution (B) to remove traces of oxygen. Place the desiccator in ice as soon as the first sample is dry.

Keep the N_2 stream on for one hour, then open the desiccator. (Exposure to the air for only a few minutes at the evaporation temperature will cause a noticeable loss of unsaturated fatty acids).

Add 0·5 ml of H_2SO_4 (A) to samples and standards, carefully moistening the walls of the test tubes, to dissolve all lipids.

Heat for 10 minutes in a waterbath at 100°C. Cool quickly to room temperature, add 5 ml of the phosphovanillin reagent (C) and measure the absorbance of the reagent blank, standards and samples, against H_2O, in suitable cells at 530 nm after exactly 30 minutes.

7.7 PROTEINS

LEVEL II

Colorimetric

Introduction

Proteins may be estimated approximately by multiplying the organic N-content (see § 5.5) by the factor 6.25. They may be estimated more accurately by the biuret reaction (Krey, 1951) or with an acetonyl-acetone reagent (Keeler, 1959). The last is 10 times as sensitive as the biuret reaction.

In the biuret method albumin is used as a standard; casein is used in the acetonylacetone method.

Agreement between the three methods is poor. Unfortunately attempts do not appear to have been made to correlate the different methods using the same standard substances.

Proteins are hydrolysed to amino acids and released from planktonic organisms (both animal and plant) by alkaline hydrolysis.

Two colorimetric methods for proteins are described here in detail.

More specific is the determination of α-amino-N (van Slyke, *et al* 1941). Amino acids may be separated quantitatively by ion exchange chromatography. They may be separated and qualitatively determined by paper chromatography (Lederer and Lederer, 1957) and by thin-layer chromatography (Stahl, 1970). See also Allen *et al*, 1974.

7.7.1 Biuret method

Principle

Biuret, $H_2NCONHCONH_2$, and all compounds containing two CONH groups at two adjacent C-atoms (or at two C-atoms bound together either by a N or by a third C-atom) form a violet coloured complex with Cu^{2+} in strongly alkaline conditions.

A correction must be made for the self colour of the reagent and for turbidity. The standard used must be specified.

Reagents

A Albumin standard solution, 1 mg ml^{-1} of albumin
Dissolve 100 mg of albumin in 100 ml of 0·2 M NaOH.

B NaOH, 2 M
Dilute 10 M NaOH, (see § 1.3.4).

C $CuSO_4$ solution, 1%
Dissolve 1·2 g of $CuSO_4.5H_2O$ in 60 ml of H_2O. Add 50 ml of (ethylene) glycol, to keep the Cu^{2+} in solution in the alkaline conditions of this procedure.

Procedure
Collect the plankton organisms on a filter (for phytoplankton a membrane filter may be used). Suspend the material for 12–24 hours in 1·0 ml of 2 M NaOH (B). (For animal material a homogenate in 0·2 M NaOH may be used. This is quicker than the extraction with NaOH.) Dilute with H_2O to 15 ml and filter through a membrane filter. Measure the absorbance against H_2O at 530 nm and at 750 nm (self colour). Add 0·5 ml of $CuSO_4$ solution (C) to 10 ml filtrate and measure the absorbance of the reagent blank, standards and samples against H_2O, in suitable cells at 530 nm and 750 nm after 30 minutes.

Prepare a calibration curve in the range 0·1–0·6 mg of albumin in 10 ml.

Calculation
Subtract the absorbance of the blank at 530 nm and 750 nm from that of the samples. Then subtract the original 'self colour' absorbance at 530 nm and 750 nm. Finally find the 'true biuret colour' by subtracting the absorbance at 750 nm from that at 530 nm.

Sensitivity
The lowest quantity of protein that can be estimated is about 30 μg per sample. Krey (1951) reports a precision of 5–10%. The accuracy depends on the type of protein used for standardization and on the type of protein in the sample, and cannot therefore be specified.

Interferences
Humic acids interfere in the case of nanoseston (Blažka 1966a).

7.7.2 Acetonylacetone reaction method

Principle
Proteins, certain amino acids and amines form pyrroles with acetonylacetone. The pyrroles form coloured complexes with p-dimethylaminobenzaldehyde (Keeler, 1959).

Reagents

A Standard protein solution, 1 mg ml^{-1}
Dissolve 100 mg of egg albumin or casein in 100 ml of 0·2 M NaOH.

B Acetonylacetone reagent
Add 1 ml of acetonylacetone to 50 ml of 0·5 M Na_2CO_3 (5 g of Na_2CO_3, or 14 g of $Na_2CO_3.10H_2O$, per 100 ml of H_2O.)

C Dimethylaminobenzaldehyde solution (Ehrlich reagent)
Dissolve 0·8 g of p-dimethylaminobenzaldehyde-HCl in a mixture of 30 ml of methanol plus 30 ml of 12 M HCl.

D Ethanol, reagent grade

Procedure
Mix 5 ml of the sample or extract (obtained as in procedure § 7.7.1) with 5 ml acetonylacetone reagent (B) in a test tube. Cover the tubes, mix, and place the tubes in a boiling water bath for 15 minutes. Cool and add 35 ml of ethanol (D) and 5 ml of Ehrlich reagent (C).

Measure the absorbance of the reagent blanks, standards and samples against H_2O in suitable cells at 530 nm after 30 minutes. Prepare a calibration curve in the range 0·05–5·0 mg of protein per sample.

All volumes may be reduced by a factor of 5, which may be useful if a 5 cm cell holding 10 ml is available. The range of application is then 0·01–1·0 mg of protein.

7.8 CHLOROPHYLL AND PHAEOPHYTIN

Several types of chlorophyll occur in plants; the most important quantitatively is usually chl *a*. Chlorophylls are easily altered (for example in even mildly acid conditions or strong light at room temperature). The products are phaeophytins. The absorption spectra of chlorophylls and phaeophytins are all slightly different, and may be used for quantitative estimations. Quantitive extraction is difficult, especially from green algae. Different solvents and temperatures should be compared.

Originally acetone was used as extractant and the extraction was carried out in the cold.

Marker (1972, 1977) and Golterman (unpublished) found that methanol, (especially 80–90% methanol at boiling point) was a better extractant.

Marker found that with some Chlorococcales acetone extracted as little as 10%. Nevertheless acetone is still used, probably because the absorption coefficient is thought to be better known. However, from several (unpublished) studies it is clear that the absorption coefficient in acetone is still in doubt. The absorbance of a methanol extract at 665 nm is always 15–20% lower than in acetone (equal concentrations). Absolute calibrations of very pure chlorophyll have not yet been published. In some circumstances acetone may be just as effective as methanol, for example when diatoms are dominant or for macrophytes.

As (commercial) methanol or ethanol tend to be slightly acid it is best to store them over solid $MgCO_3$.

LEVEL I

7.8.1 Colorimetric

Principle
In cases where there is little phaeophytin present, and most of the chlorophyll is chl *a*, an approximate estimate of chl *a* may be got from the absorbance at

665 nm of a methanol or acetone extract. The results are expressed, as if all the chlorophyll were chl *a*, as chl *a* equivalent. The solvent yielding the highest absorbance should be used.

Apparatus
Colorimeter and filter (see § 7.8.2).

Reagents

Acetone, ethanol or methanol (reagent grade) stored over 0·1% $MgCO_3$ (solid). Filters retain some water. The organic solvent concentration should be such that, after adding it to the filter, the final concentration of organic solvent is 80–90%.

Procedure
Filter off the particulate matter from a known volume of water sample. Either glass fibre or membrane filters may be used. Roll the filter, put it in a centrifuge tube, and add a sufficient measured volume of (ice cold) solvent to cover the filter. Cover the tubes, and put them in a refrigerator for 24 hours if acetone is used. If the solvent is ethanol or methanol heat to near boiling in a very dim light. Decant the solvent, and if turbid, centrifuge. Measure the absorbance in a suitable cell at 665 nm.

Calculation

Let x = concentration of chl *a* equivalent in the *solvent* [μg ml^{-1}]
then for acetone $x = 11{\cdot}9\ E_1$
for ethanol and methanol $x = 13{\cdot}9\ E_1$
where E_1 = absorbance at 665 nm in 1 cm cell

The factors 11·9 and 13·9 are from Talling and Driver (1963).

LEVEL II

7.8.2 Colorimetric or spectrophotometric

Principle
The absorption spectrum of chlorophyll has a maximum in acetone at 663 nm. Chlorophyll can be converted to phaeophytin by the addition of an acid, which removes Mg from the chlorophyll molecule. Both molecules occur in natural conditions. Phaeophytin also absorbs light at 663 nm, but less strongly than the same concentration of chlorophyll. From the decrease in absorbance when the sample is acidified the amount of chlorophyll can be calculated but only in aqueous organic solvents. In many older methods the acid is too concentrated (Marker, 1977).

An approximate correction for other coloured compounds, and for turbidity, can be made by subtracting the absorbance at 750 nm (where chlorophyll and phaeophytin absorb an insignificant amount of light).

Accurate measurement of the concentration of chlorophylls is extremely difficult. The method described here is easy and rapid in use but the results should be interpreted with great care. For example the two most common chlorophylls, *a* and *b*, both absorb light at 663 nm and the absorption coefficients are not accurately known (see Table 7.2). For further information see Report of SCOR, 1966, and Brown, 1968. Macrophytes must be homogenised before extraction.

Apparatus

(1) Filter assembly for 47 mm diameter filters including filtering flask and device for producing a vacuum.
(2) Whatman GF/C or Gelman A filters.
(3) Grinding apparatus. It is convenient to have this driven electrically. Samples may even be ground by hand if macroscopic plants are present.
(4) 15–20 ml screwcap test tubes.

If no spectrophotometer is available a colorimeter and suitable optical filter may be employed. If possible the device should be standardised by comparing a range of absorbance readings on a spectrophotometer with the colorimetric readings. The optical filter should have suitable cutoff properties so that it transmits only light within the principal chlorophyll absorption region centred about 663 nm. Suitable filters are Corning 2–58, Wratten no. 26 and Schott RG/2. The correction at 750 nm (described in the Procedure), cannot be made with this apparatus.

Reagents

A $MgCO_3$ (reagent grade, powdered)
Make a suspension in H_2O in a plastic bottle.

B Acetone, ethanol or methanol (reagent grade) stored over 0·1% solid $MgCO_3$.
See § 7.8.1 and note B.

C HCl, 0·2 M See § 1.3.4.

Procedure

(a) FOR MACROSCOPIC PLANTS

Place a filter on the tower and apply vacuum. Deposit a film of $MgCO_3$ on the filter by filtering a suspension of it in H_2O. (The $MgCO_3$ increases retention of particles on the filter). Pour a suitable measured volume of the fresh water sample through the filter and suck dry. Remove the filter, fold in half so that the side having precipitate is not exposed and drop it into the grinding tube. Add 2 ml of 90% acetone, insert the pestle and grind to release pigments from the cells: 30 seconds at 1000 rpm should be sufficient. Remove the pestle and wash with 2 ml of 90% acetone, catching the washings in the grinding tube. Wash the contents of the grinding tube into the screwcap test tube and make up to a known volume, generally 10 ml. Centrifuge for 30 seconds. Measure the absorbance of the supernatant in a suitable cell, with a ground stopper or lid at 663 nm and at 750 nm. Pour the contents of the cuvette back into the screwcap test tube, add 0·1 ml of 4 M HCl (C), and recentrifuge. Measure the extinction again at 663 nm and 750 nm.

(b) FOR MICROSCOPIC PLANTS

Use the extraction procedure § 7.8.1. Then follow the measurement and acidification stages of § 7.8.2 (a) above.

Calculation

Let U' = absorbance of unacidified extract corrected for absorbance at 750 nm and for 1 cm path length

A' = absorbance of acidified extract corrected in the same way as U'

then $$U' = \frac{U_{663} - U_{750}}{l}$$

$$A' = \frac{A_{663} - A_{750}}{l}$$

where l = path length of the spectrophotometer cell. Subscripts 663 and 750 refer to measurements at 663 nm and 750 nm

Let x = concentration of chlorophyll + phaeophytrin [mg l^{-1}]

then $$x = \frac{10^6\, U'\, V_e}{K_c V_s}$$

where V_e = volume of extract [ml]

V_s = volume of sample [ml]

K_c = absorption coefficient of chlorophyll: 91 (see note E and Table 7.2)

Let P_c = absorbance of chlorophyll

P_p = absorbance of phaeophytin

then $P_c = 2{\cdot}43\,(U' - A')$

$P_p = U' - P_c$

Let y = concentration of chlorophyll [mg l^{-1}]

z = concentration of phaeophytin [mg l^{-1}]

then $$y = \frac{10^6\, P_c\, V_e}{K_c\, V_s} = \frac{2{\cdot}43 \times 10^6\,(U' - A')\, V_e}{K_c\, V_s}$$

$$z = \frac{10^6\, P_p\, V_e}{K_p\, V_s} = \frac{10^6\,(2{\cdot}43\, U' - 1{\cdot}43\, A')\, V_e}{K_p\, V_s}$$

where K_p = absorption coefficient of phaeophytin: 55 (see note E)

100 μg l^{-1} is a fairly high concentration in lake water.

Notes

A. If samples are to be stored for some reason, they should be stored in a darkened desiccator at around 0°C. It is, however, emphasised that best results are obtained when samples are analysed immediately. In no circumstances should extracts be stored for longer than 24 hours.

B. On the whole, methanol is a better extractant than 90% acetone. The absorption coefficients are not well known however and changes in the pigment structure occur rapidly in methanol or acetone, especially when these solvents are slightly acid. This can be avoided by storing them over solid $MgCO_3$.

C. In the acidification step, 100% chlorophyll solution will yield a ratio U'/A' of 1·7 or slightly higher. If the extract contains nothing but pheaophytin, the ratio will be 1·0 (see Fig. 7.3).

D. From Fig. 7.3, U'/A' can be converted directly to percentage as chlorophyll or phaeophytin.

E. There is about 25% difference in the absorption coefficients given by different workers. Because of the nature of the methods used for obtaining these values, one might recommend using the highest absorption coefficient for chl *a*. The recommended values for chlorophyll and phaeophytin are 91 and 55 respectively. The value of 91 is not consistent with the calibration value 11·9 given in § 7.8.1. This difference can be explained by the presence of chl *b*. The recommended values for chl *a* in methanol are 15–20% lower so that 78 and 47 are recommended (Talling and Driver, 1963). Much data has been published however using the value 65 which is close to the value given for phaeophytin.

F. In the calculation above 2·43 derives from 1·7/(1·7–1·0).

LEVEL III

7.8.3. Fluorimetric

Chlorophyll can also be measured by its fluorescence. The method may be used after extraction and is highly sensitive. It is useful when small amounts of pigment are to be estimated. The high sensitivity is obtained by using very efficient photomultipliers. Two filters are required. The first is blue, and is placed in the excitation beam. The second is red and allows the passage of fluoresced red light to the photomultiplier tube.

For details see Yentsch and Menzel (1963), Holm-Hansen *et al.* (1965).

The fluorimetric determination can also be used for live cells (so called 'in vivo'

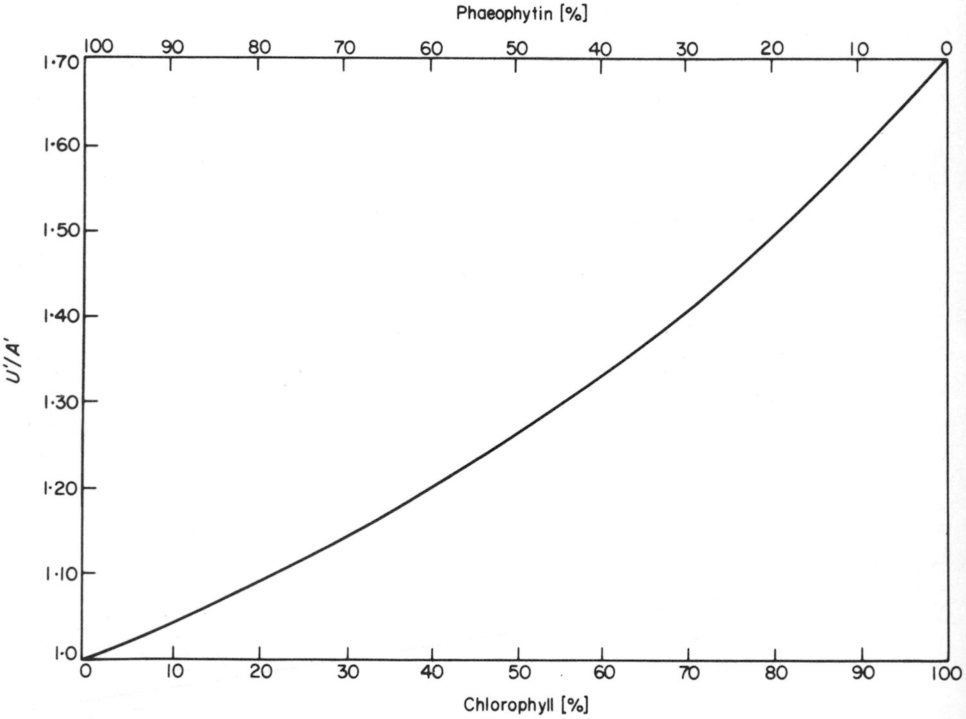

Figure 7.3. Relationship between percentages of chlorophyll and phaeophytin and the absorbance ratio (see text).

Table 7.2. Absorption coefficient (K) for chl *a* in aqueous acetone.

The absorption coefficient $K_c = \frac{\log_{10} I/I_0}{lC}$

where l is the light path in centimeters and C is the concentration in g l^{-1}.

Reference		Absorption coefficient (K) for wavelength 665 nm	664–663 nm	Acetone concentration
Zscheile	1934 a, b	65·0	68·5	90%
MacKinney	1940	84·0	—	100%
MacKinney	1941	81·0	82·0	80%
Zscheile *et al.*	1942	—	82·0	80%
Richards and Thompson	1952	66·7	71·0	90%
Vernon	1960	90·8	92·6	100%
			91·1	90%
Parsons and Strickland	1963	89·0	90·0	90%

measurements) by pumping the lake water directly into a fluorometer. Lorenzen (1966) describes this technique. The measurements were made using a Turner model III fluorometer equipped with a red-sensitive photomultiplier (R 136) and a blue fluorescent lamp with Corning CS 5–60 and CS 2–64 filters.

A large difference has been observed between measurements on extracts and on live cells. Kiefer (1973 a, b) and Loftus and Seliger (1975) have established that much of this could be attributed to the irradiance experienced by the algal cells, to nutrient stress and to interspecific differences.

The fluorescence of live blue green algae has been found to be appreciably lower than for other algal groups (Heaney, 1978). This will be one factor contributing to variability in freshwater. Fee (1976) and George (1976) conclude that the method is useful for showing local differences within a lake.

7.9 CHLOROPLAST PIGMENTS

LEVEL III

Separation by thin-layer chromatography

Principle

The substances to be separated are placed in a spot or streak near the edge of a thin layer of adsorbent on a glass plate. As in paper chromatography, a solvent is made to pass across the adsorbent layer. If the solvent and adsorbent are correctly chosen, the substances to be separated move across the plate at different rates,

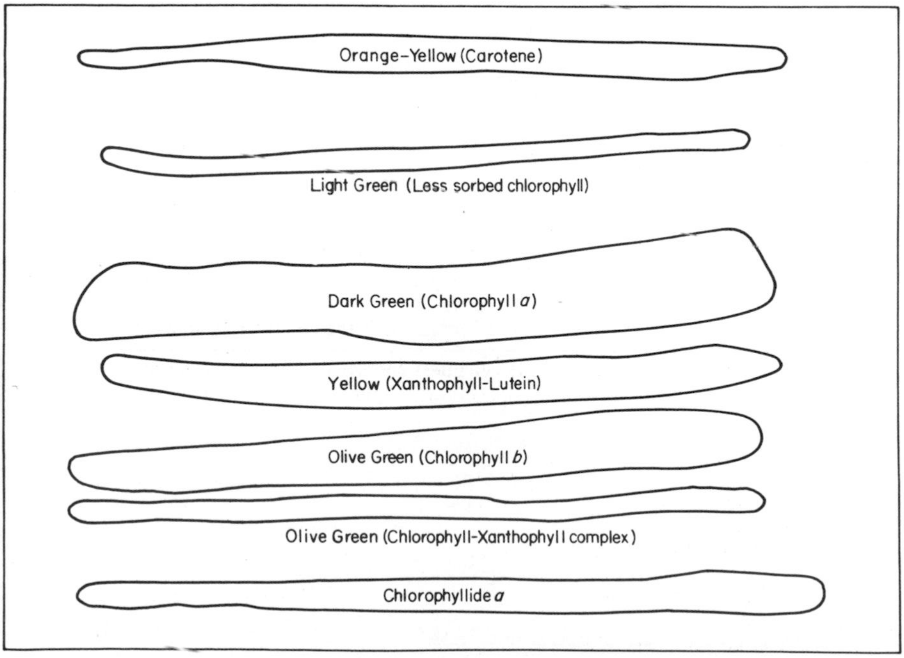

Figure 7.4. Thin-layer chromatogram of extract from a green alga.

because of differences in their chemical structure, and come to form discrete spots or streaks.

The method is rapid—a complete run may take only 30 minutes. It is also possible to use larger amounts of material than in paper chromatography.

For application of the technique to pigment separations see Madgwick (1966), Bollinger *et al.* (1965), Stahl (1970), Bauer (1952) and Riley and Wilson (1967).

Procedure (after Riley and Wilson, 1967).
Extract the pigments with 10 ml 90% acetone or methanol. Mix the extract with 15 ml of diethyl ether. (Caution. Ether is highly inflammable.) Extract the mixture three times with 25 ml of 10% (w/v) NaCl solution. Evaporate the ether at room temperature under N_2 and dry the residue in vacuo. Dissolve the pigment mixture in 100 μl of acetone and apply as a line 1 cm from one edge of a 20 × 20 cm glass plate coated with a 0·25 mm layer of Silica Gel G (Merck). A simple method of plate preparation is illustrated in Fig. 7.6. Develop the plate with a solvent containing petroleum ether (b.p. 60–80°C), ethyl acetate and diethylamine (58:30:12 by volume). Remove the coloured bands from the plate and elute the pigments with acetone for further examination. Figs 7.4 and 7.5 show typical thin-layer chromatograms of algal extracts.

Identification of pigments can be carried out using R_f values, absorption spectra and co-chromatography with authentic pigments on a thin-layer plate coated with glucose (Madgwick, 1966). The eluted pigments may also be rechromatographed on glucose (Madg ›ick, 1966) if there is reason to suppose that they are not homogeneous (e.g. dinoxanthine and diadinoxanthin).

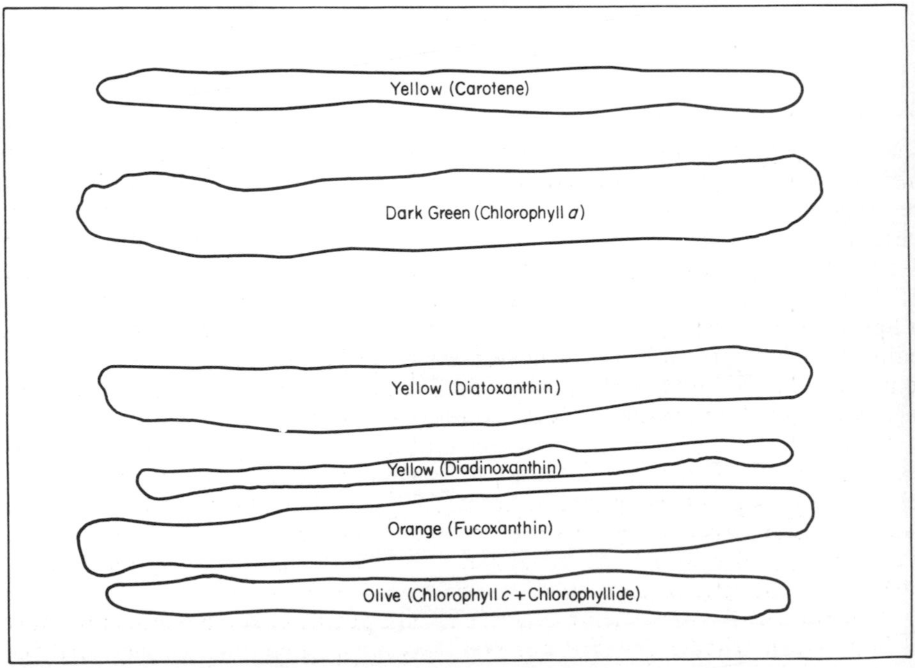

Figure 7.5. Thin-layer chromatogram of extract from a brown alga.

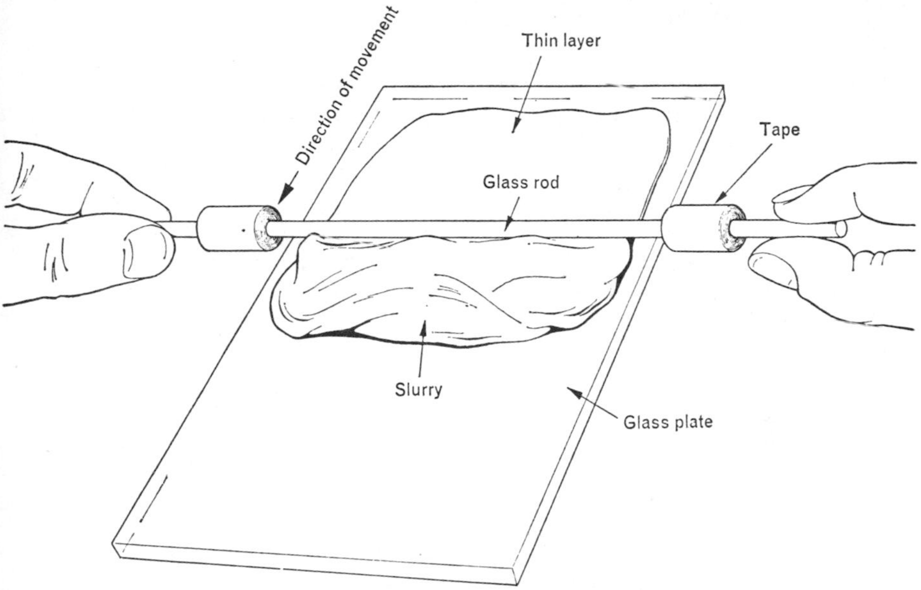

Figure 7.6. Glass rod for applying adsorbent.

7.10 PESTICIDES AND RELATED COMPOUNDS

LEVEL III

Gas liquid chromatography

The most frequent analyses carried out for pesticides in freshwater concern certain fairly persistent organochlorine residues, which can be found almost anywhere if the level of detection is sufficiently reduced. Because of the low concentration of these residues in water they are detected by gas liquid chromatography using an electron capture detector. They include residues of the DDT group, dieldrin and HCH isomers. They are stable, relatively insoluble in water and highly soluble in lipids. They therefore accumulate rapidly in animal tissue and are adsorbed on sediments. If the presence of organochlorine residues is suspected in water it may be useful to make preliminary analyses on fish or sediments. For some pesticides, negative results here indicate that the water is also free.

Other pesticides and herbicides may be found in water. Organophosphorus pesticides are widely used, but they are mainly non-persistent and are only detectable in the immediate vicinity of a recent discharge. The phenoxy herbicides are also short lived, with restricted toxicity; their detection requires the preparation of derivatives if the same GLC technique is used as for the other organochlorines. Carbamates and triazine herbicides are difficult to detect.

A manual of this size cannot attempt to give details of the many complicated methods which exist for pesticide analyses. Anyone wishing to make such analyses should consult an expert in this field; it is easy to obtain worthless results. Because the organochlorine residues are those most frequently examined in water, some information is given here on their extraction and analysis.

Storage and Treatment of Water Samples

Avoid the use of plastic containers and filter papers. Use all-glass equipment and filter the water if necessary through pre-washed glass-wool filters. Add 4 ml of H_2SO_4 (1 + 1) per litre as a preservative. Store the sample in a refrigerator, preferably in the dark. Wash all containers with pure solvent before use.

Large samples (10–20 litres) are usually required because of the extremely low concentration of pesticide residues in water. Rinse the sample containers with a water-miscible solvent such as acetone when the sample has been removed, to ensure that no material remains on the container walls. Combine the rinse with the sample.

Principle

The organochlorine residues are extracted either by continuous solvent extraction of the water, or by passage of the water through absorbent columns of various types of materials (e.g. resins and polyurethane foam). The choice of solvent and material for the column depends on the pesticide being examined. For details of extractions with resins see Burnham *et al* (1972), Coburn *et al* (1977), Junk *et al* (1974), Leoni *et al* (1975), Musty and Nickless (1974a); for extractions with polyurethane foam see Bedford (1974), Musty and Nickless (1974b) and Uthe *et al* (1972). The residues are removed from the columns by elution with organic solvents.

The organic extract or eluate is then evaporated to a small volume (Breidenbach *et al*, 1964; Holden and Marsden, 1966) and the concentrate subjected to a clean-up to remove unwanted materials (Cooke, 1972; Holden and Marsden, 1969; Pesticide Analytical Manual, 1968). This clean-up involves either a liquid-liquid partition or passage through a column of adsorbent, such as alumina, silica or Florisil. A preliminary separation of residues can be obtained by this chromatographic clean-up (Armour and Burke, 1970; Holden and Marsden, 1969). The eluates are then examined by GLC (Goerlitz and Brown,1972).

Confirmation of the identity of the residue is obtained in part by the relative retention time of the GLC peak, and by the eluate in which it is removed from a chromatographic column. Identification may also be carried out by chemical techniques (Cochrane and Chau, 1971), or by mass spectrometry.

Thin-layer chromatography has been used for the examination of organochlorine residues (Allen *et al*, 1974; Goodenkauf and Erdei, 1964; Kovacs, 1963) but the technique is only useful when several micrograms of material are available.

Reagents

High quality reagents must be used throughout. Analytical grade solvents normally require redistillation and adsorbents require pre-treatment with solvents before use.

Interferences

Blanks should be used to check for possible contamination. An extremely high standard of laboratory cleanliness is essential to avoid contamination from the laboratory itself.

7.11 DISSOLVED VITAMIN B_{12}, THIAMINE AND BIOTIN

LEVEL III

Bioassay

Principle
As the concentrations of vitamins in fresh water are generally very low, they can only be estimated by bioassay. In bioassay the growth, in the water sample, of an organism that needs the compound in question for its growth, is compared with the growth of the same organism in solutions of known concentrations of the vitamin.

Detailed procedures for the bioassay of dissolved vitamin B_{12}, thiamine, and biotin in fresh water systems are available in Kavanagh (1963). More recently, rapid and sensitive algal bioassay methods for these vitamins in sea water have been developed [Carlucci and Silbernagel (1966a), (1966b), (1967a), 1967b)] and it is recommended that these be adapted for fresh water assays. *Cyclotella nana* (13–1), *Monochrysis lutheri*, and *Amphidinium carteri* are used to assay vitamin B_{12}, thiamine, and biotin, respectively.

Outline of procedure for seawater
Samples are collected as aseptically as possible, sterilized by passage through a membrane filter, and frozen at −20°C until assayed. Prior to assay the samples are thawed, refiltered if any question exists as to sterility, supplemented with sterile nutrients, dispensed into appropriate vessels, inoculated with the test alga, and incubated in the light. After 46 hours (96 hours in the biotin assay) $Na^{14}CO_3$ is added and then the cultures are incubated for two more hours. The algal cells are collected on a filter and ^{14}C uptake measured. The ^{14}C uptake is proportional to vitamin concentration in ecologically significant ranges. A standard curve is prepared using known additions to a vitamin-free (charcoal-treated) seawater. A known amount of the vitamin is also added to a portion of each sample to serve as standard addition. From the amount of vitamin recovered from standard addition the inhibitory properties of the sample can be calculated. If facilities for using ^{14}C are not available, the assay cultures are incubated for two or three more days and cell numbers determined. After a sufficient incubation period, cell numbers are proportional to vitamin concentration.

Range of application
The ranges of vitamin concentration (in seawater) which can be assayed are: vitamin B_{12}, 0·05–3·0 ng l^{-1}; thiamine, 2–35 ng l^{-1} and biotin, 0·2–6 ng l^{-1} (1 ng $= 10^{-9}$g).

Notes
The analyst should refer to the original papers for the detailed procedures for these assays. The modification necessary for fresh water analysis is to dilute the sample with NaCl supplemented, charcoal-treated seawater (artificial or natural seawater may be used) so that the final concentration of salts in the diluted sample is 28–35‰. The analyst may find it more useful and practical to modify these assays so that fresh water algae can be used as the test organisms. A list of algae requiring vitamins has been compiled by Lewin (1961).

CHAPTER 8

DISSOLVED GASES

8.1 OXYGEN

General principles

The determination of O_2 may be carried out with an O_2-detecting electrode or by a titrimetric method. The electrode methods are most useful when a large number of measurements are required in the field, and possess the special advantage that the collection of separate samples in bottles is not necessary, because an electrode may be lowered over the side of a boat to give direct measurements.

The titrimetric methods are mainly based on the technique described by Winkler. See § 8.1.2.

It is possible to buy field kits for estimating the I_2 colour produced in the Winkler method. The manufacturer's instructions must be followed carefully. Apparatus of this type is supplied by British Drug Houses and by Hach Chemical Company.

Additional methods have been described by Roskam and Langen (1963), Montgomery *et al.* (1964), Knowles and Lowden (1953), Hart (1967), and Bryan *et al.* (1976).

LEVEL I

8.1.1. Titrimetric

Principle

See § 8.1.2, Level II

Apparatus, sampling, reagents

See § 8.1.2, Level II.

The level II method (§ 8.1.2) can easily be simplified and performed in the field using starch as indicator. If the titration has to be postponed, the samples should be kept in the alkaline stage and preferably be kept cool, dark and sealed from the air.

The reagents A and B of procedure § 8.1.2 can then be kept in the laboratory.

LEVEL II

8.1.2. Titrimetric

Principle

O_2 combines with $Mn(OH)_2$ forming higher hydroxides which, on subsequent acidification in the presence of I^-, liberate I_2 in an amount equivalent to the original dissolved O_2 content of the sample. The I_2 is then determined by titration with $Na_2S_2O_3$. Interference by NO_2^- up to 0·36 mM (= 5 mg l^{-1} NO_2–N) is eliminated by the use of NaN_3 (sodium azide).

The recommended procedure employs a reagent devised by Pomeroy and Kirschman (1945), which contains a much higher concentration of NaI than the reagent used formerly. The advantages are that errors due to the volatilization of I_2 and to interference by organic matter are reduced, that the hydroxides dissolve more readily, and that the starch end point is sharper.

Apparatus

The bottles should be of good quality, with narrow necks and well fitting ground glass stoppers. In normal use bottles are kept clean by the acidic iodine solution produced during the Winkler procedure and require no treatment apart from rinsing with H_2O. Contaminated bottles must be discarded. It is easier to exclude gas bubbles if the stopper end is cut obliquely.

Sampling

Special care is needed in sampling for dissolved O_2. For accurate work a displacement sampler is necessary. The lower end of the outlet tube of the sampler is placed just above the bottom of the Winkler bottle. The contents of the sample bottle are displaced three times, before the sample is collected. For very shallow water the sample may be taken in a syringe and analysed by a small-scale technique such as that of Fox and Wingfield (1938). The bottle or sample container must always be completely filled and reagents or preservatives must be added immediately after sampling. Sampling from shallow waters and errors in sampling in general are discussed by Montgomery and Cockburn (1964).

Reagents

A $\frac{1}{6}KIO_3$, 0·100 M

Dissolve 3·567 g of KIO_3 (A.R., dried 105°C) in H_2O and dilute to 1000 ml. Dilute further as required.

B_1 $Na_2S_2O_3$, approximately 0·025 M

Dissolve 6·2 g of $Na_2S_2O_3.5H_2O$, in H_2O. Add a pellet of NaOH or 1 g of Na_2CO_3 and dilute to 1 litre. Store in a brown bottle.

B_2 $Na_2S_2O_3$, 0·0125 M, or 0·0025 M (= 12·5 mM or 2·5 mM)

Dilute the 0·025 M $Na_2S_2O_3$ solution to the required strength and store in a brown bottle. The strength required will depend on the size of the sample: for 250 ml bottles (and for the BOD test) 0·0125 M $Na_2S_2O_3$ is normally used. For smaller samples, 0·0025 M may be used without loss of accuracy.

For standardization, which should be repeated frequently (at least each week), pipette 10·00 ml of 0·025 M KIO_3 solution into a conical flask containing 100 ml of H_2O and add 1 ml of alkaline iodide-azide solution (D), followed by 2 ml of H_3PO_4 (C). Mix *thoroughly* then titrate with the 0·0125 M $Na_2S_2O_3$, adding 2 ml of starch solution (F) or about 0·5 g of starch-urea just before the end point. If the strength of $Na_2S_2O_3$ is other than 0·0125 M adjust the concentration of standard KIO_3 and alter the volumes of H_2O and reagents in proportion, so that an actual dissolved O_2 titration is simulated as closely as possible in the standardization.

C H_3PO_4, (s.g. 1·75)

D Alkaline iodide-azide solution

Dissolve 400 g of NaOH in 560 ml of H_2O, add 900 g of NaI (A.R.) and keep the solution hot until the NaI has dissolved. Cool the solution and dilute to 1 litre. Decant or filter, if necessary, after standing overnight. No I_2 should be liberated when 1 ml is diluted to 50 ml and acidified. For work in which the highest accuracy is not required, a reagent containing 500 g of NaOH and 140 g of NaI (or 150 g of KI) per litre may be used.

If more than 10 μg of NO_2–N is present mix 1 litre of alkaline NaI reagent with 300 ml of a 2·5% NaN_3 solution.

For work of the highest accuracy standardise in the absence of NaN_3 because this substance interferes slightly in the reaction between IO_3^- and I^- giving an error of about 1%. As the NaN_3 also interferes in the samples it follows that in waters with more than 10 μg of NO_2–N the highest accuracy cannot be obtained.

If only impure (I_2-containing) NaI is available, dissolve 900 g in 560 ml of H_2O, add a few drops of CH_3COOH and stir with 1 g of Zn dust until the solution is colourless. Filter and add 400 g of NaOH without delay, then cool and dilute to 1 litre.

E $MnSO_4.5H_2O$, 50%

Dissolve 500 g of $MnSO_4.5H_2O$ in H_2O, filter if necessary, and make up to 1 litre. No iodine should be liberated when 1 ml of reagent is added to 50 ml of acidified iodide solution.

F Starch indicator, 1%

Disperse 1 g of starch in 100 ml of water and warm to 80°–90°C. Stir well, allow to cool, and add 0·1 g salicylic acid. Alternatively use powdered starch-urea complex ('cold water soluble starch'), which can be purchased ready for use ('Thiodene').

G Polyviol MO5/140 (Wacker–Chemie GmbH or Bush Beach; alternative to starch as end point indicator)

This is a polyvinyl alcohol. The colour change is from orange to colourless.

Procedure

The whole procedure must be carried out away from direct sunlight.

To the sample add 1 ml of $MnSO_4$ solution (E) well below the neck of the bottle and 1 ml of alkaline iodide-azide solution (D) at the surface (the fine points of pipettes may be cut off so that they empty more quickly or hypodermic syringes may be used). Incline the bottle, replace the stopper carefully so as to avoid inclusion of air bubbles and thoroughly mix the contents by inverting and rotating the bottle several times for about ten seconds. When the precipitate has settled to the lower third of the bottle, repeat the mixing and then allow the precipitate to settle completely leaving a clear supernatant.

In the absence of organic matter the estimation of the dissolved O_2 by acidification and titration may be postponed at this stage, provided that the bottles are kept in the dark. It is, of course, vital that air be excluded from the bottle during the period between precipitation and acidification; this should be assured if bottles with well-fitting stoppers are used, but even then the period should not exceed a few hours. Bottles may be kept cool by immersion in cold water.

Add 2 ml of H_3PO_4 (C). Replace the stopper and mix the contents thoroughly by rotation (a bubble of CO_2 may form at this stage, but this is not important). Normally the precipitate will dissolve almost instantaneously; if it does not, allow to stand for a few minutes and repeat the mixing. In any case mix the contents of the bottle again immediately before measurement.

Measure into a conical flask the whole sample or a suitable aliquot (say 50 ml) of the solution and immediately titrate the I_2 with $Na_2S_2O_3$ (B_1 or B_2) solution, using as indicator 2 ml of starch solution (F) or about 0·5 g of starch-urea, added only towards the end of the titration. For very precise work an amperometric or dead stop method (§ 9.13) may be used to detect the end point. See § 8.1.4.

Notes

A. The procedure applies to sample bottles of nominal capacity 125 ml. If bottles of very different size are used (e.g. 250 ml) the amounts of reagents must be adjusted in proportion, and the calculation must be adjusted accordingly.

B. I_2 is volatile and therefore the titration must be carried out as quickly as possible and with the minimum of exposure to the air. For very low concentrations of dissolved O_2, polyviol MO5/140 (G) may be used instead of starch.

Interferences

A modification of the Winkler method suitable for the determination of dissolved O_2 in swamp water has been described by Beadle (1958). Another modification which has been used for waters with a high organic content is the bromination procedure of Alsterberg (1926). Success with Alsterberg's procedure requires the use of an alkaline NaI reagent with a high concentration of NaI (as already described here). The sodium salicylate reagent, which destroys the excess of Br_2, must be freshly prepared and colourless; it is best prepared from crystalline sodium salicylate. Br_2 destroys NO_2^-, hence NaN_3 is not required. Rebsdorf (1966) found that the Alsterberg modification gave the same results as the Winkler method in unpolluted and higher results in polluted waters. A comprehensive review of modifications of the Winkler method is given by Legler (1972) who pointed out that several modifications lead to no improvement because they need more time, and introduce new sources of error.

Calculation

LEVELS I AND II

Allowance is made in the formula given below for the slight displacement of sample by the $MnSO_4$ and alkaline NaI reagents, which contain very little dissolved O_2.

Let $x = O_2$ concentration in water [mg l^{-1}]

Then
$$x = \frac{8{\cdot}0\, C_b\, V_b}{V_a\,(V_f - 2{\cdot}0)/V_f}$$

where C_b = concentration of reagent B [mmol l^{-1}]
V_b = volume of reagent B [ml]
V_a = volume of aliquot taken for titration [ml]
V_f = bottle volume, with stopper in place [ml]

If the whole contents of the bottle are titrated then

$$x = \frac{8{\cdot}0\, C_b\, V_b}{(V_f - 2{\cdot}0)}$$

The bottle volume V_f should be measured—mostly easily by weighing—with an accuracy and precision of 0·1%.

Solubility and saturation depend on temperature and atmospheric pressure.

Let z = saturation [%]

Then $z = 100x/S_{T,P}$

where $S_{T,P}$ = solubility of O_2 in water at temperature T and pressure P (Table 8.1). For saline water see Carpenter (1966).

Alternatively use one of the nomograms Figure 8.1 or 8.2.

Figure 8.1 must be followed by a small calculation to include pressure. Figure 8.2 includes pressure but is (as printed) less accurate at low oxygen concentration.

Table 8.1. Solubility of O_2 in water in equilibrium with air at 760 mm Hg pressure and 100% relative humidity (from EAWAG 1973) Units: mg l^{-1}.

°C	0·0	·1	·2	·3	·4	·5	·6	·7	·8	·9
0	14·60	14·56	14·52	14·48	14·44	14·40	14·36	14·33	14·29	14·25
1	14·21	14·17	14·13	14·09	14·05	14·02	13·98	13·94	13·90	13·87
2	13·83	13·79	13·75	13·72	13·68	13·64	13·61	13·57	13·54	13·50
3	13·46	13·43	13·39	13·36	13·32	13·29	13·25	13·22	13·18	13·15
4	13·11	13·08	13·04	13·01	12·98	12·94	12·91	12·88	12·84	12·81
5	12·78	12·74	12·71	12·68	12·64	12·61	12·58	12·55	12·52	12·48
6	12·45	12·42	12·39	12·36	12·33	12·29	12·26	12·23	12·20	12·17
7	12·14	12·11	12·08	12·05	12·02	11·99	11·96	11·93	11·90	11·87
8	11·84	11·81	11·78	11·76	11·73	11·70	11·67	11·64	11·61	11·58
9	11·56	11·53	11·50	11·47	11·44	11·42	11·39	11·36	11·34	11·31
10	11·28	11·25	11·23	11·20	11·17	11·15	11·12	11·10	11·07	11·04
11	11·02	10·99	10·97	10·94	10·91	10·89	10·86	10·84	10·81	10·79
12	10·76	10·74	10·72	10·69	10·67	10·64	10·62	10·59	10·57	10·55
13	10·52	10·50	10·47	10·45	10·43	10·40	10·38	10·36	10·34	10·31
14	10·29	10·27	10·24	10·22	10·20	10·18	10·15	10·13	10·11	10·09
15	10·07	10·04	10·02	10·00	9·98	9·96	9·94	9·92	9·89	9·87
16	9·85	9·83	9·81	9·79	9·77	9·75	9·73	9·71	9·69	9·67
17	9·65	9·63	9·61	9·59	9·57	9·55	9·53	9·51	9·49	9·47
18	9·45	9·43	9·41	9·39	9·37	9·36	9·34	9·32	9·30	9·28
19	9·26	9·24	9·23	9·21	9·19	9·17	9·15	9·13	9·12	9·10
20	9·08	9·06	9·05	9·03	9·01	8·99	8·98	8·96	8·94	8·92
21	8·91	8·89	8·87	8·86	8·84	8·82	8·81	8·79	8·77	8·76
22	8·74	8·72	8·71	8·69	8·67	8·66	8·64	8·63	8·61	8·59
23	8·58	8·56	8·55	8·53	8·51	8·50	8·48	8·47	8·45	8·44
24	8·42	8·41	8·39	8·38	8·36	8·35	8·33	8·32	8·30	8·29
25	8·27	8·26	8·24	8·23	8·21	8·20	8·18	8·17	8·16	8·14
26	8·13	8·11	8·10	8·08	8·07	8·06	8·04	8·03	8·01	8·00
27	7·99	7·97	7·96	7·94	7·93	7·92	7·90	7·89	7·88	7·86
28	7·85	7·84	7·82	7·81	7·80	7·78	7·77	7·76	7·74	7·73
29	7·72	7·70	7·69	7·68	7·66	7·65	7·64	7·63	7·61	7·60
30	7·59	7·57	7·56	7·55	7·54	7·52	7·51	7·50	7·49	7·47

If the barometric pressure at the time of sampling is not 760 mm Hg then the saturation at the actual pressure will differ from those given in the Table. 760 mm Hg = 1 atm = 101325 Pa (= Nm^{-2}), where Pa commemorates Pascal.

Let y = solubility of O_2 in water at the measured pressure (mg l^{-1})

then $$y = \frac{SP_p}{101325} = \frac{SP_m}{760}$$

where P_p = measured pressure [Pa]

P_m = measured pressure [mm Hg]

S = solubility of O_2 in water at 760 mm Hg pressure (from Table 8.1)

It is therefore necessary in accurate work, to note the barometer reading at the time of sampling. For less accurate work the average barometric pressure may be estimated from the altitude (Table 8.2).

Table 8.2. Variation of mean atmospheric pressure with altitude (Dussart and Francis-Boeuf 1949)

Altitude (m)	Average atmospheric pressure (mm Hg)	Factor	Altitude (m)	Average atmospheric pressure (mm Hg)	Factor
0	760	1·00	1300	647	1·17
100	750	1·01	1400	639	1·19
200	741	1·03	1500	631	1·20
300	732	1·04	1600	623	1·22
400	723	1·05	1700	615	1·24
500	714	1·06	1800	608	1·25
600	705	1·08	1900	601	1·26
700	696	1·09	2000	594	1·28
800	687	1·11	2100	587	1·30
900	679	1·12	2200	580	1·31
1000	671	1·13	2300	573	1·33
1100	663	1·15	2400	566	1·34
1200	655	1·16	2500	560	1·36

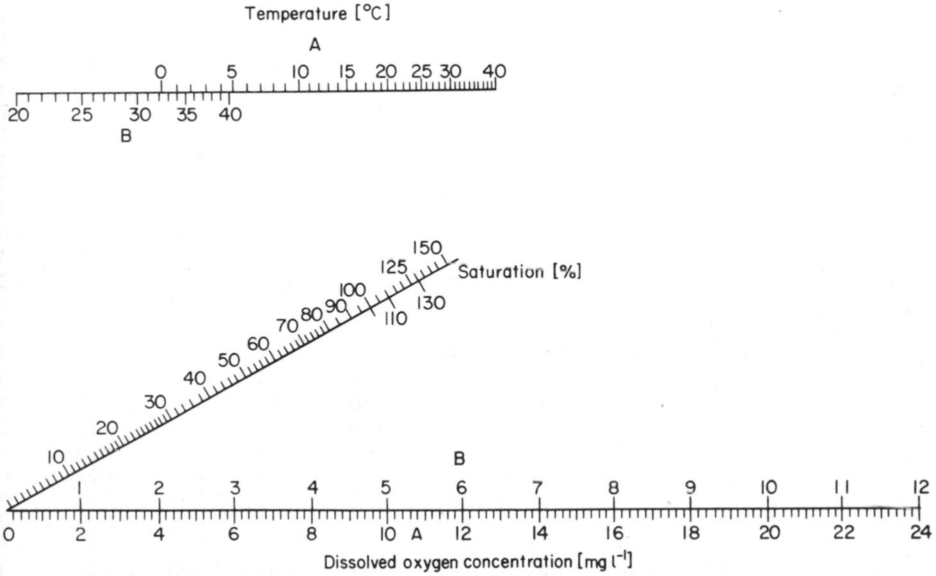

Figure 8.1. Nomogram for interconverting O_2 concentration and % saturation in fresh water.
Draw a line connecting the temperature with the oxygen concentration, using the A scales for temperature and dissolved O_2. The percentage saturation is then read from the point at which the line crosses the central scale. For temperature above 20°C improved accuracy may be obtained by the use of the B scales, but these scales may only be used when the percentage saturation does not exceed 100. Finally, correct the percentage saturation to standard pressure.

Let $x =$ saturation at standard pressure [%]

then $$x = \frac{101325\ S_n}{P_p} = \frac{760\ S_n}{P_m}$$

where S_n = saturation from the nomogram [%]
Other symbols as for Table 8.1.

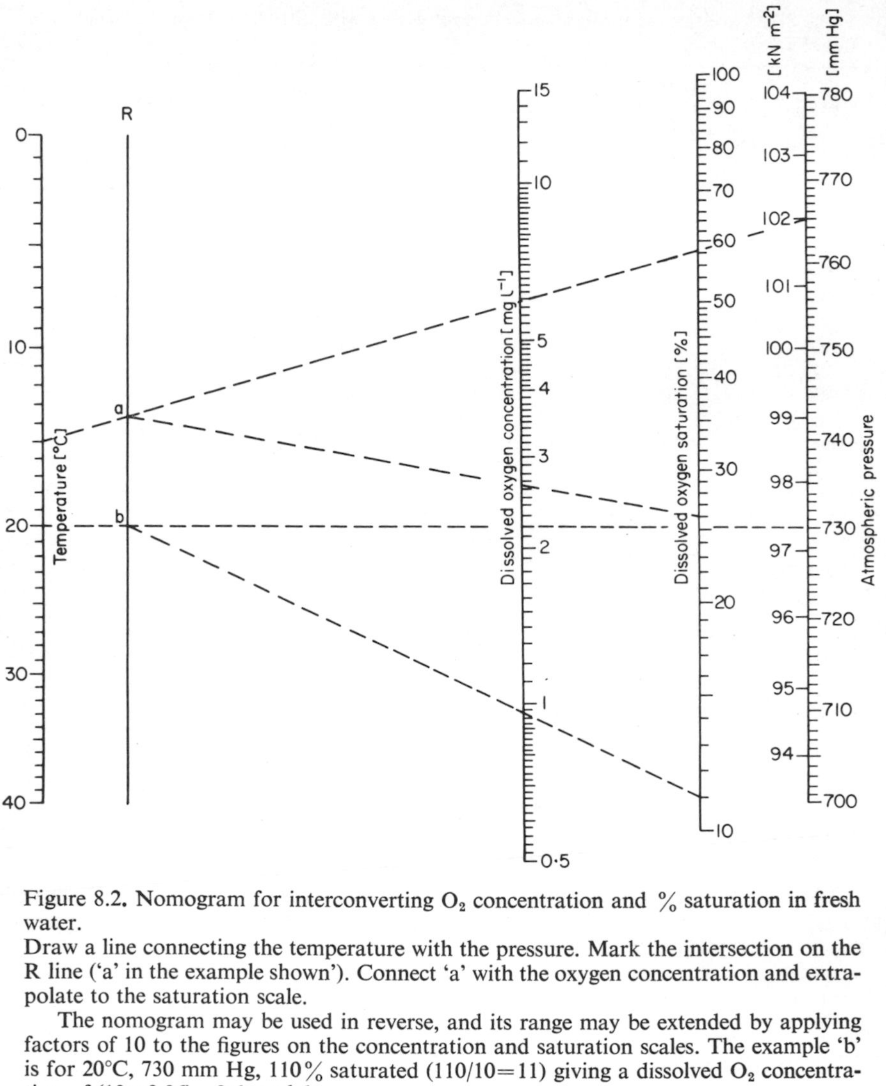

Figure 8.2. Nomogram for interconverting O_2 concentration and % saturation in fresh water.
Draw a line connecting the temperature with the pressure. Mark the intersection on the R line ('a' in the example shown'). Connect 'a' with the oxygen concentration and extrapolate to the saturation scale.

The nomogram may be used in reverse, and its range may be extended by applying factors of 10 to the figures on the concentration and saturation scales. The example 'b' is for 20°C, 730 mm Hg, 110% saturated (110/10=11) giving a dissolved O_2 concentration of $(10 \times 0{\cdot}96) = 9{\cdot}6$ mg l^{-1}.
Figure reproduced by permission of Dr. I. C. Hart, HMSO and Pergamon Press.

LEVEL II AND LEVEL III

8.1.3. 'Electrode' methods

Principle
O_2 passes by diffusion to a reducing electrode—uncovered in the dropping mercury polarographic method, membrane-covered in the Clark cell—or to a membrane covered galvanic cell ('Mackereth electrode').

The amount of O_2 reduced in unit time can be made proportional to the concentration of O_2 in the water, and the resulting electrical current or potential can be amplified and displayed. The usefulness of the uncovered type of cell is limited because the electrode system may be easily poisoned. The electrode is in direct contact with the water and surface-active compounds, as well as other suspended material, frequently adsorb on its surface, particularly from wastewaters. This disadvantage is avoided in membrane covered 'electrodes', which allow diffusion of O_2 but prevent that of electroactive and surface active interferents. The two types of membrane covered cell are similar in operating characteristics. In the voltammetric type (Clark cell) an appropriate emf source is needed, while the galvanic type (Mackereth cell) is basically an oxygen energized cell. All that is needed with the galvanic cell oxygen analyzer is a low impedance galvanometer or microammeter. For recording, an inexpensive galvanometer-type recorder is adequate in most cases though a potentiometric recorder may also be used.

All these cells *consume* O_2, so the water around them must be kept moving. In detail the principles of these methods are very different. For further information see Mackereth (1964), Mancy and Okun (1960), Mancy and Westgarth (1962), Mancy et al. (1962).

H_2O containing a known concentration of O_2 is required for calibration of the electrode. Either the O_2 content of the calibration water may be found by the method § 8.1.2 or H_2O at a known temperature may be bubbled with air at a known barometric pressure. The electrical measurement is made while the solution is being stirred but not bubbled. These processes are alternated until there is no further change in reading. The percentage saturation is then 100 and the concentration of O_2 may be found from Table 8.1.

Apparatus

The construction of a membrane-covered cell has been described by Mackereth (1964), and a temperature-compensation circuit has been published by Briggs and Viney (1964). Temperature compensated cells may be obtained commercially. A dropping mercury device for field use has been described by Føyn (1955, 1967).

The Clark polarographic cell (Pt-Ag) requires an external polarising potential to generate a relatively small current. This in turn requires amplification. This makes the earlier versions of apparatus less reliable in the field but excellent in the laboratory; but modern integrated circuit technology is removing this limitation. The Clark cell uses less O_2 than the Mackereth type. The original Mackereth probe was not temperature compensated but current commercially available apparatus (for example EIL) is.

Membrane-covered cell instruments may be bought from

The Lakeland Instruments Co.	Simac Ltd.
Electronic Instruments Ltd.	Beckmann
Honeywell, Inc.	Precision Instruments Ltd.
Yellow Springs Instrument Co., Inc.	Uniprobe Instruments Ltd.

Some of these instruments require daily calibration. Where automatic temperature compensation is not provided in the instrument, the temperature of the water must be noted and a correction applied according to the manufacturer's instructions.

Procedure

Use the apparatus according to the author's or manufacturer's instructions. The instructions regarding response time and turbulence should be noted carefully.

Clean the electrode (for example with 1 M HCl) as often as there is visible contamination or when the performance becomes sluggish. Calibrate the electrode frequently—at least once a day.

8.1.4. 'Dead-stop' titrimetric

The 'dead stop' technique (see § 9.13) is recommended for this level. This method can easily be automated. Excess thiosulphate is used and back titrated.

No further level III techniques are recommended but attention is drawn to the highly precise modification of the Winkler method described by Carpenter (1965) and by Knowles and Lowden (1953).

The membrane electrode technique may be used with continuous recording, but 'aufwuchs' is difficult to avoid and may lead to serious errors.

8.2 METHANE

LEVEL II

8.2.1. Combustible-gas indicator method

Principle

The partial pressure of CH_4 in the gas phase (which is in equilibrium with the solution) may be determined with a combustible gas indicator. The combustible gas is oxidised catalytically on a heated Pt filament, which is part of a Wheatstone bridge network. The electrical resistance of the filament depends on the heat of oxidation of the gas. Frequent calibration is necessary (Rossum, Villarruz and Wade, 1950). The method is relatively simple and quick.

Precision, accuracy and sensitivity

The sensitivity is about 0·2 mg l^{-1} CH_4. The precision may be 2%. The accuracy varies between 2% and 10% depending on the calibration curve.

LEVEL III

8.2.2. Combustion-absorption method

Principle

CH_4 is transferred to the gas phase and catalytically oxidised to CO_2. The CO_2 is estimated by the volume change on absorption in KOH. The methods of procedures § 7.1. 4–7 may also be used. For further details see Larson (1938) and Mullen (1955).

Precision, accuracy and sensitivity

The sensitivity is 2 mg l^{-1} of CH_4. The precision and accuracy depend on the gasometric operations and may be less than 5%.

LEVEL III

8.2.3. Gas chromatographic method

The determination of CH_4 in the gas chromatograph is relatively simple. The gas must be transferred from the dissolved phase to the gas phase with an inert gas, which serves as a carrier gas as well. The apparatus must be calibrated frequently.

8.3 OXYGEN, NITROGEN AND TOTAL CARBON DIOXIDE

LEVEL III

8.3.1. Gas chromatographic

The use of a gas chromatographic method for O_2, N_2 and total CO_2 has been described by Jeffery and Kipping (1964), Park (1965), Park and Catalfomo (1964), Swinnerton, Linnenbom and Cheek (1962a and b, 1964).

Montgomery and Quarmby (1966) describe a method for extracting the gases O_2, N_2 and CO_2 from water and their measurement by gas chromatography.

The method is relatively simple and rapid. It has the advantage that the three gases are determined in one sample.

High sensitivity has been obtained by Stainton (1973). He placed 25 ml of sample in a syringe, then sucked in a small measured volume of helium and a measured volume of acid. The mixed gas volume was then injected into the gas chromatograph. Calibration was made with $NaHCO_3$.

It is essential to avoid cavitation resulting from excessive pressure reduction if a syringe is filled too quickly.

General note

If a 'once through' method is used it should be demonstrated that the dissolved gases are completely evolved into a sufficiently small volume of carrier gas.

8.4 HYDROGEN, CARBON DIOXIDE AND METHANE

LEVEL III

Gas chromatographic (tentative)

Ray (1954) has separated the components of a mixture of H_2, CH_4, CO_2, ethylene and ethane by gas-solid chromatography. Although the method has not yet been applied in limnochemistry there seem to be no obvious difficulties in applying it to fresh water.

CHAPTER 9

ELECTROCHEMICAL PROPERTIES OF WATER

THEORETICAL ADDENDUM TO CHAPTER 3

9.1 CONDUCTIVITY

The conductivity of a solution is a measure of its capacity to convey an electric current. Conductivity is related to the nature and concentration of ionized substances present in the solution and to the temperature of the solution. Even pure H_2O has a measurable conductivity. After absorbing CO_2 from the air, and other substances from the glass in which H_2O is kept, the conductivity of distilled water is normally about 1 μS cm^{-1} = $10^{-6}\ \Omega^{-1}\ cm^{-1}$.

Kohlrausch has measured a value as low as 0·05 μS cm^{-1}. This indicates that water acts as a very weak binary electrolyte according to the following equation:

$$H_2O \rightleftharpoons H^+ + OH^- \tag{9.1}$$

From this conductivity it can be calculated that the concentration of H^+ and OH^- is about 10^{-7} M (= 10^{-7} mol l^{-1}).

By more accurate measurements—using electrode potentials—a value of exactly 10^{-7} has been found at 24°C ($0{\cdot}83 \times 10^{-7}$ at 20°C).

According to the law of homogenous equilibrium

$$[H^+][OH^-] = K[H_2O] = K_w \tag{9.2}$$

where K_w is known as the ionization constant of water. This 'constant' depends upon temperature.

The ionization constant—like the dissociation constants to be discussed later—is strictly speaking defined in terms of the 'activity', rather than the concentration.

The activity is the 'apparent' concentration in non ideal (i.e. more concentrated) solutions, where interactions between ions play a role. Even in fresh water the activity coefficients can be markedly less than 1·0, but they are not considered further here.

Substituting the value $0{\cdot}83 \times 10^{-7}$ for $[H^+]$ and $[OH^-]$ in equation (9.2), the value of K_w becomes $0{\cdot}68 \times 10^{-14}$ at 20°C. K_w increases with the temperature because the degree of ionization is greater at higher temperatures. The conductivity depends therefore on the temperature. It is inconvenient to write the concentration of H^+ as $0{\cdot}93 \times 10^{-7}$ (or 0·000,000,093) M. The terms pH and pK are therefore used. They are defined as:

$$pH \equiv -\log_{10}[H^+] \qquad pK \equiv -\log_{10}K$$

where K is the ionization constant.

The variation of pK_w and pH with temperature is shown in Table 9.1.

It should be noted that the concentrations of *both* H^+ and OH^- increase with temperature, in contrast to what happens after the addition of an acid to water as is described in § 9.2.

The conductivity of solutions is the sum of the conductivities of the individual species of positive and negative ions. The molar conductivity of an ion is determined partly by its mobility in the electric field. (The mobility is expressed in cm sec^{-1} per V cm^{-1}, thus in $cm^2\ V^{-1}\ sec^{-1}$, in which V is the potential difference.)

The mobility of the ions depends on the temperature so the conductivity of a solution depends on the temperature. The conductivity also depends on the concentration and the degree of ionization of the salts involved. This causes a non-linear relationship between conductivity and concentration (see Table 3.3).

The conductivity gives no indication of the nature of the substances in solution but any increase or decrease in their concentration will be reflected in a corresponding increase or decrease in conductivity.

The concentration of dissolved ionic matter [mg l^{-1}] in a fresh water sample may be roughly estimated by multiplying the conductivity by an empirical factor, varying from 0·55 to 0·9 depending on the soluble compounds in the particular water (for KCl, 0·51; for NaCl, 0·48; for $Ca(HCO_3)_2$, 1·6). A rough approximation to the total concentration of charged solutes (mmol l^{-1}) may be obtained by multiplying the conductivity in $\mu S\ cm^{-1}$ by 0·01.

Table 9.1. Variation of pK_w and pH of pure H_2O with temperature. (From Handbook of Chemistry and Physics.)
Note that some authorities report pK_w = 14·000 at 22°C.

Temperature (°C)	pK_w	pH
0°	14·943	7·47
15°	14·346	7·17
20°	14·167	7·08
24°	14·000	7·00
25°	13·997	7·00
30°	13·833	6·92

Conductance can be used for conductometric titrations of, for instance' alkalinity and chloride. In soft water the conductometric method has advantages over the alkalinity titration method using a pH meter. In all cases, but especially at low Cl^- concentrations, the conductometric method has advantages over the potentiometric method, as in the conductometric method use is made of all the readings, except those close to the end points, to construct the straight lines used in estimating the end point. This makes use of most of the measurements, in contrast to the electrometric end point determination, in which it is mainly those points close to the end point which are used. The conductometric method is therefore especially useful for waters with a low HCO_3^- or Cl^- concentration. For a typical titration curve, see Fig. 9.3.

9.2 pH

In a neutral solution the concentration of H^+ is equal to that of OH^-. If an acid is added, the $[H^+]$ is increased and consequently the $[OH^-]$ is decreased according to equation 9.2.

The pH of neutral water at 24°C is 7·0 since the $[H^+]$ is 10^{-7} M. The addition for instance of 1 mmol l^{-1} of a strong acid makes the $[H^+]$ in the water 10^{-3} M = 10^{-3} mmol l^{-1}: the pH becomes 3.

Weak acids are not completely ionized in water: an equilibrium is established between dissociated and undissociated molecules, the degree of ionization depending on the dilution.

Acetic acid for instance is dissociated according to:

$$CH_3COOH \rightleftharpoons CH_3COO^- + H^+ \tag{9.3}$$

Applying the law of homogeneous equilibrium

$$K = \frac{[CH_3COO^-][H^+]}{[CH_3COOH]} \tag{9.4}$$

If the degree of ionization in reaction (9.3) is called α

$$\alpha \equiv \frac{[CH_3COO^-]}{[CH_3COO^-] + [CH_3COOH]}$$

and V is the volume in which 1 mole of the acid is dissolved then

$$[CH_3COO^-] = [H^+] = \frac{\alpha}{V}$$

and

$$K = \frac{\alpha^2}{(1 - \alpha)V} \tag{9.5}$$

(This equation summarizes Ostwald's dilution law.)

If V = 1000 litre (i.e. 0·001 M), α = 0·127, as $K = 18{\cdot}4 \times 10^{-6}$ for acetic acid; the H^+ concentration is then $0{\cdot}127 \times 10^{-3}$: the pH = 3·9.

For very weak acids (pK > 5) the dissociation is often so small that $(1 - \alpha) \simeq 1$. Equation (9.5) then becomes:

$$K = \frac{\alpha^2}{V} \quad \text{thus} \quad \alpha = \sqrt{(KV)} \tag{9.6}$$

Equation (9.5) does not hold for acids with $K > 0{\cdot}5 \times 10^{-2}$ i.e. pK > 2·3. HCl for instance has, for practical purposes, no ionization constant, because it dissociates almost completely.

9.3 BUFFER SOLUTIONS

When a salt of a weak acid (for instance CH_3COONa) is added to a solution of the acid itself the following equilibria exist:

$$HAc \rightleftharpoons H^+ + Ac^-$$
$$NaAc \rightleftharpoons Na^+ + Ac^-$$

Since HAc is a weak acid, only a small proportion ionizes. On the other hand, nearly all the NaAc exists in the ionic form. Substituting in equation (9.4)

$$K = \frac{[NaAc]_{added}[H^+]}{[HAc]_{added}}$$

and thus

$$pH = pK + \log \frac{[NaAc]}{[HAc]} \tag{9.7}$$

If a small quantity of an acid is added to this solution the H^+ of the acid immediately combine with the acetate-ions and form undissociated acetic acid molecules leaving the $[H^+]$ practically unaltered. More exactly, or when a large quantity of acid is added (for instance x mol l^{-1}) the pH becomes

$$pH = pK + \log \frac{[NaAc]_{added} - [x]}{[HAc]_{added} + [x]} \tag{9.8}$$

From this formula it can be seen that to cause a decrease of 1 unit in pH

$$\frac{[NaAc] - [x]}{[HAc] + [x]} \text{ must be } \frac{1}{10} \cdot \frac{[NaAc]}{[HAc]}$$

Assuming a concentration of NaAc and HAc of 0·1 M (pH = 4·75), x must be about 0·08 mol l^{-1} H^+ to decrease the pH by 1 unit to pH = 3·75. Without buffer the pH would have dropped to 1·1. If a base is added the OH^- of the base combine with the H^+ from the acetic acid. More acetic acid then ionizes to replace the H^+ until equilibrium is established again at not much below the original H^+ concentration. By varying the relative proportions of sodium acetate and acetic acid, solutions of any desired initial pH (within certain limits) may be prepared. Standard buffer solutions of a definite pH are therefore prepared from mixtures of NaAc and HAc or from other weak acids and their salts, for instance H_3PO_4, or H_2CO_3 with Na_2CO_3. The last of these is the buffer system of most alkaline lakes. A particularly useful series of buffer solutions, with pH in the biologically important range, can be made from mixtures of Na_2HPO_4 and NaH_2PO_4.

Buffer solutions of high concentration are more stable over long periods of time than those containing small amounts of acids or bases which may very easily be affected by absorption of CO_2 from the air or by solution of alkali from the glass container, or even by H^+ or OH^- from the ionization of any indicator which is added. Useful buffer mixtures are described in § 3.2.4.

9.4 INDICATORS

One of the most common methods of determining pH is to add an appropriate indicator to the unknown solution and to compare the resulting colour with that of buffer solutions of known pH value containing the same indicator. The indicators selected for this purpose are for the most part weak organic acids or bases which are able to exist in two forms possessing different colours. Being weak acids or bases, the law of homogeneous equilibrium applies to them just as it does to any other weak acid or base. The equilibrium may be indicated by the following equations:

$$HIn' \rightleftharpoons HIn \rightleftharpoons H^+ + In^-$$

$$\frac{[H^+][In^-]}{[HIn']} = K$$

HIn has the same colour as In^-, its anion. HIn′ is the tautomeric form of the undissociated molecule (HIn), and has a different colour from HIn and In^-. Very little HIn is present. The colour of the indicator is determined by the ratio

$[In^-]/[HIn']$. Since this ratio multiplied by the $[H^+]$ is equal to a constant, the ratio and, therefore, the colour of the indicator itself is directly related to the $[H^+]$.

It has been found that the human eye can detect no further visible colour change when the ratio of $[In^-]/[HIn']$ becomes greater than approximately 91/9 or less than 9/91. That is, the visible colour change occurs within a hundred-fold change in the concentration of either In^- or HIn'. There is a corresponding hundred-fold change in $[H^+]$. The usable colour range, then, of indicators in general corresponds to a range of 2 pH units.

Although the range of any one indicator is thus limited, there is fortunately considerable variation in the ranges of the numerous indicators available, so that almost any $[H^+]$ may be determined by the proper selection of indicator. A list of some indicators is given in § 3.2.4.

9.5 INFLUENCE OF TEMPERATURE ON THE pH OF WATER SAMPLES

The most important buffer system in natural waters with a pH > 6·5 is the bicarbonate–carbonate system. The pH of this system can be calculated from

$$\text{pH} = \text{p}K + \log \frac{[HCO_3^-]}{[CO_2]}, \text{ see equation (9.7)}$$

Temperature has an influence on the pH because it affects the dissociation coefficients of acids, and the solubility of CO_2. When the pH of a water sample is measured at a different temperature from that at which it was collected, care must therefore be taken that the samples are stored in closed bottles without air bubbles. The pH can also be measured at the 'in situ' temperature and corrected by calculation to a standard temperature for all samples. As a rule the pH decreases by about 0·1 pH unit with a temperature increase of 20°C.

When using colour indicators the temperature of the colour indicator itself must also be taken into account, as the dissociation coefficients of the indicators change as well. A particular colour shade therefore does not indicate the same pH at all temperatures. Most pH meters have a temperature compensating mechanism, (either a potentiometer, or a thermistor placed in the solution). It is important to realise that this corrects only for the temperature sensitivity of the electrodes and that the 'true' pH of the solutions at the prevailing temperature is measured.

Additional complications are met if different parts of the electrode system are at different temperatures, and this situation is best avoided if possible.

It will be clear that no general advice can be given about temperature correction. The analyst must decide whether he requires the pH at the in situ temperature or at laboratory temperature, or at some other temperature. He must also decide how accurately he needs to know the pH, and then decide the appropriate procedure for himself.

9.6 THE CARBONATE, BICARBONATE, CARBONIC ACID SYSTEM

Many natural waters contain CO_3^{2-}, HCO_3^-, and free CO_2, all in equilibrium with each other. As an example consider a solution of $NaHCO_3$ in H_2O. (In most natural waters the dominant cation is Ca^{2+}, but the theory is more complicated than for Na^+. For a complete treatment see Stumm and Morgan, 1970).

This salt, when added to the water dissociates into Na^+ and HCO_3^-. The latter reacts with the H^+, which originates from the dissociation of H_2O, to form H_2CO_3. In turn, H_2CO_3 dissociates into CO_2 (in solution) and H_2O, the CO_2 being in equilibrium with the CO_2 from the air. These reactions are rapid so that within a short time a series of equilibria are all being satisfied simultaneously.

The reactions are the following:

$$NaHCO_3 \rightarrow Na^+ + HCO_3^- \quad (9.9)$$

$$H_2O \rightleftharpoons H^+ + OH^- \quad (9.10)$$

$$HCO_3^- + H^+ \rightleftharpoons H_2CO_3 \quad (9.11_1)$$

$$H_2CO_3 \rightleftharpoons H_2O + CO_2 \quad (9.11_2)$$

$$HCO_3^- \rightleftharpoons H^+ + CO_3^{2-} \quad (9.12)$$

The equilibrium constant for reactions (9.11_1) and (9.11_2) together is called K_1, the apparent dissociation constant for the first ionization step of carbonic acid. The dissociation constant of (9.12) is K_2. At 25°C $K_1 = 4{\cdot}30 \times 10^{-7}$ and $K_2 = 5{\cdot}61 \times 10^{-11}$. Both are temperature dependent (see Table 9.2).

Table 9.2. Temperature dependance of the first (K_1) and second (K_2) ionization constant of carbonic acid and of the ionization product K_w of water. (Harned and Owen, 1958; Bates, 1964.)

Temp. (°C)	0	5	10	15	20	25
pK_1	6·58	6·52	6·46	6·42	6·38	6·35
pK_2	10·63	10·56	10·49	10·43	10·38	10·33
pK_w	14·94	14·73	14·54	14·35	14·17	14·00

If a change in the system (e.g. by changing the pH) is induced, more of the HCO_3^- will follow equation 9.11_1 than will follow 9.12 because $K_1 \gg K_2$. As a reaction 9.12 is therefore of little quantitative importance in most lake waters. It is useful to sum reactions 9.9, 9.10 and 9.11:

$$NaHCO_3 \rightleftharpoons Na^+ + OH^- + CO_2 \quad (9.13)$$

It can be seen from (9.13) that when $NaHCO_3$ is dissolved in H_2O the solutions are slightly alkaline (pH = 8·3 for 10^{-3} M $NaHCO_3$). When normal air containing 320 vpm CO_2 at 10^5 Pa (= 1 atm) pressure at 20°C is in equilibrium with H_2O then there is 0·5 mg l^{-1} CO_2 in solution. It often happens that if 9.11_2 is to be satisfied there must be a higher concentration than this.

The excess of CO_2 formed will—until the system is in equilibrium with the air—escape. The disappearance of the CO_2 and thus the decrease in $[H^+]$ due to this process allows the $[OH^-]$ to increase. Accordingly to reaction (9.12) CO_3^{2-} then increases.

As K_2 is much smaller than K_1 the solution becomes more strongly alkaline.

The overall reaction, the summation of reactions (9.9, 9.10, 9.11 and 9.12) is:

$$2NaHCO_3 \rightleftharpoons 2Na^+ + CO_3^{2-} + H_2O + CO_2 \quad (9.14)$$

But this reaction should not suggest that 1 mole of CO_3^{2-} is formed if 1 mole of CO_2 is withdrawn from the system because the consequent pH change will cause more CO_3^{2-} to be formed.

Reaction (9.14) will take place until the equilibrium between HCO_3^-, CO_3^{2-} and 'equilibrium' CO_2 has established itself. ['Equilibrium' CO_2 refers to CO_2 in equilibrium with HCO_3^- and CO_3^{2-}; it is not, in general, in equilibrium with air]. The equilibrium position depends on K_1 and K_2 and on the pH of the solution. Figure 9.1 gives the percentage of HCO_3^-, CO_3^{2-} and CO_2 present in a solution in relation to the pH.

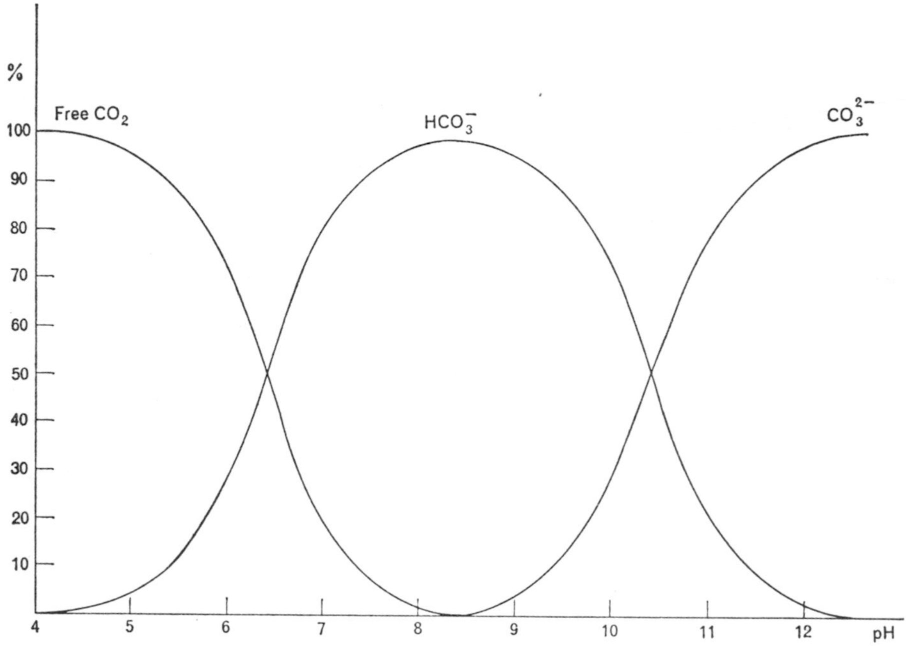

Figure 9.1. Relation between pH and % of total 'CO_2' as free CO_2, HCO_3^- and CO_3^{2-}. From Schmitt (1955).

During photosynthesis the water becomes more alkaline in accordance with reaction (9.14) owing to CO_2 uptake by algae.

In Table 9.3 are shown the amounts of *free* CO_2, calculated after derivation from equation (9.7) by

$$pH = pK + \log \frac{[HCO_3^-]}{[CO_2]},$$

the pH values, and the *changes of the pH with time*, of some $NaHCO_3$ solutions. The pH of a fresh solution depends partly on the CO_2 already present. This effect is more marked in very dilute solutions of $NaHCO_3$.

In most natural waters, where Ca^{2+} is the most common cation, the solubility product of $CaCO_3$ determines the amount of Ca^{2+} and CO_3^{2-} (and thus the HCO_3^- concentration) that can be in solution together. (Solubility product, which is a constant at a given temperature, is obtained as $[Ca^{2+}] \times [CO_3^{2-}]$ in a saturated solution). The dissolution of $CaCO_3$ (limestone) in nature can be described as follows. The $CaCO_3$ comes in contact with H_2O containing CO_2. Some $CaCO_3$

Table 9.3. pH values measured and concentration of CO_2 calculated in $NaHCO_3$ solutions.

	0 hours		after 24 hours	
$NaHCO_3$	pH	CO_2 mg l^{-1}	pH	CO_2 mg l^{-1}
0·001 M in unboiled H_2O	7·2	6	7·94	1·1
in boiled H_2O	8·1	0·8	8·1	0·8
0·003 M in boiled H_2O	8·35	1·1	8·60	0·8
0·010 M in boiled H_2O	8·44	3·7	8·95	1·1

will react and will go into solution as Ca^{2+} and HCO_3^- ions. The pH of the water will increase and CO_3^{2-} in solution will be formed in a pH dependent equilibrium with HCO_3^-. If or when the product of $[Ca^{2+}]$ times $[CO_3^{2-}]$ is equal to the solubility product the reaction will stop; the solution is then saturated with respect to $CaCO_3$. Theoretically, in water in equilibrium with the air, the maximum concentration of dissolved $CaCO_3$ is 0·12 mM. The concentration of $Ca(HCO_3)_2$ is then about 1 mM. For further discussion see Schmitt (1955) and Weber and Stumm (1963).

In many cases, however, metastable conditions persist for a long time with much higher Ca^{2+} concentrations (Hutchinson, 1957, page 670).

The solubility of $CaCO_3$ can also be increased by an extra amount of free CO_2. If more CO_2 is present than is in equilibrium with the air (in groundwaters, or in hypolimnia) much more $CaCO_3$ will dissolve, because this extra CO_2 keeps the pH low, thus causing a lower concentration of CO_3^{2-}. If this super-saturated water becomes exposed to air the excess CO_2 will escape causing some $CaCO_3$ to precipitate. Tillmans and Heublein (1912) have given an approximate relation between dissolved $Ca(HCO_3)_2$ and the amount of free CO_2 necessary to keep the $Ca(HCO_3)_2$ in solution (Table 9.4 and Fig. 9.2). This applies only to waters with a pH lower than 8·3 because the concentration of free CO_2 is negligible at greater pH values. These solutions are, however, not in equilibrium with the air.

Table 9.4. Relation between $CaCO_3$ concentration and free CO_2.

Concentration $\frac{1}{2}$ $CaCO_3$ in water [mmol l^{-1}]	Concentration of equilibrium CO_2 [mg l^{-1}]
1	0·6
2	2·5
3	6·5
4	15·9

The amount of free CO_2 present in excess of the equilibrium CO_2 is sometimes called 'aggressive' CO_2 as it is able to react with alkaline carbonates or metals. It can be present only in waters not in equilibrium with the air, which renders its determination extremely difficult. Some indication can be found by comparing the existing pH with the equilibrium pH of the same solution. Titration with Na_2CO_3 or NaOH to an end point of pH 8·4 gives the quantity of 'free CO_2' (= 'aggressive' + 'equilibrium'). In practice this is a difficult titration since air must be excluded, but at the same time CO_2 must not be removed.

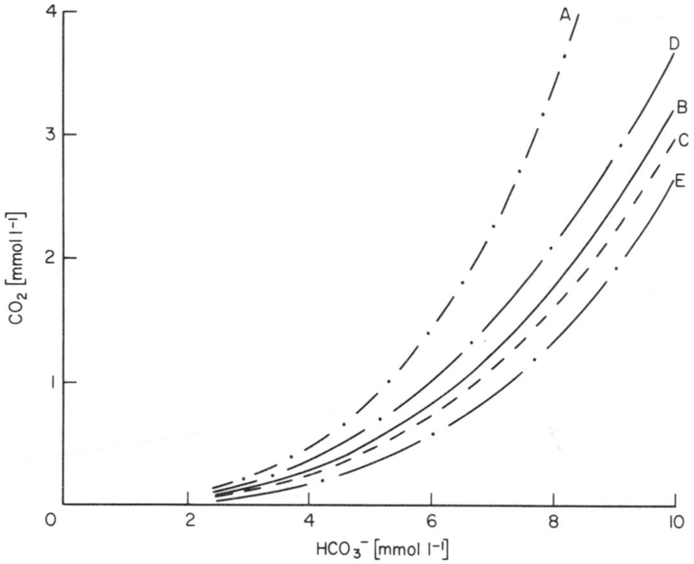

Figure 9.2. Free CO_2 necessary to keep $Ca(HCO_3)_2$ in solution at 10°C, after Tillmans and Heublein (1912) from Kleijn (1972). A uncorrected, B to E corrected, for mutual influence of ions. B for low concentrations of Ca^{2+} and HCO_3^-, and ionic strength of other ions less than 5 mmol l^{-1}. C as B, but for ionic strength of other ions greater than 5 mmol l^{-1}. D as B, but $[\frac{1}{2}\ Ca^{2+}] = [HCO_3^-] + 1$ mmol l^{-1}. E as B, but $[\frac{1}{2}\ Ca^{2+}] = [HCO_3^-] - 1$ mmol l^{-1}.

9.7 TITRATION OF THE CARBONATE–BICARBONATE SYSTEM; ALKALINITY

A solution of $NaHCO_3$ has a pH > 7 and can be neutralized with HCl.

$$Na^+ + HCO_3^- + H^+ + Cl^- \rightarrow Na^+ + Cl^- + CO_2 + H_2O$$

The change of pH as HCl is added can be seen in Fig. 9.3A$_1$. A measure of the 'alkalinity' is given by the amount of HCl used to neutralize the alkaline reaction of the water. It is usually expressed in mmol l^{-1}. At the end point the pH is not 7, but is, due to the H_2CO_3 still present, slightly lower. The titration can be carried out by detecting the end point with the colour change of an indicator with an indicator range below 6, for instance with methyl red (pH range 4·4–6·0). The titration can also be performed under N_2. The liberated CO_2 is partly removed and the titration curve follows more closely the pattern of a NaOH titration. This sharpens the end point, but does not alter its position. See Fig. 9.3A$_2$.

As the mobility of HCO_3^- is less than that of Cl^- the conductivity of the water increases linearly up to the end point. Past the end point the increase of conductivity per unit of titrant added will be greater (but again constant) per quantity of HCl added. The titration can therefore be carried out with a conductometric end point determination (see Fig. 9.3B).

The alkalinity of a water is determined by titration with strong acid. Bicarbonate ions are not the only ones which cause the alkalinity. The other main contributors are CO_3^{2-} and OH^-, and in exceptional cases even $SiO(OH)_3^-$, $H_2BO_3^-$, NH_4^+, HS^-, organic anions, and colloidal or suspended $CaCO_3$. It is possible to calculate the concentration of total CO_2 (= free CO_2 + HCO_3^- + CO_3^{2-}) from the pH and the alkalinity of the water in question, if the alkalinity is caused only by HCO_3^-, CO_3^{2-} and OH^-.

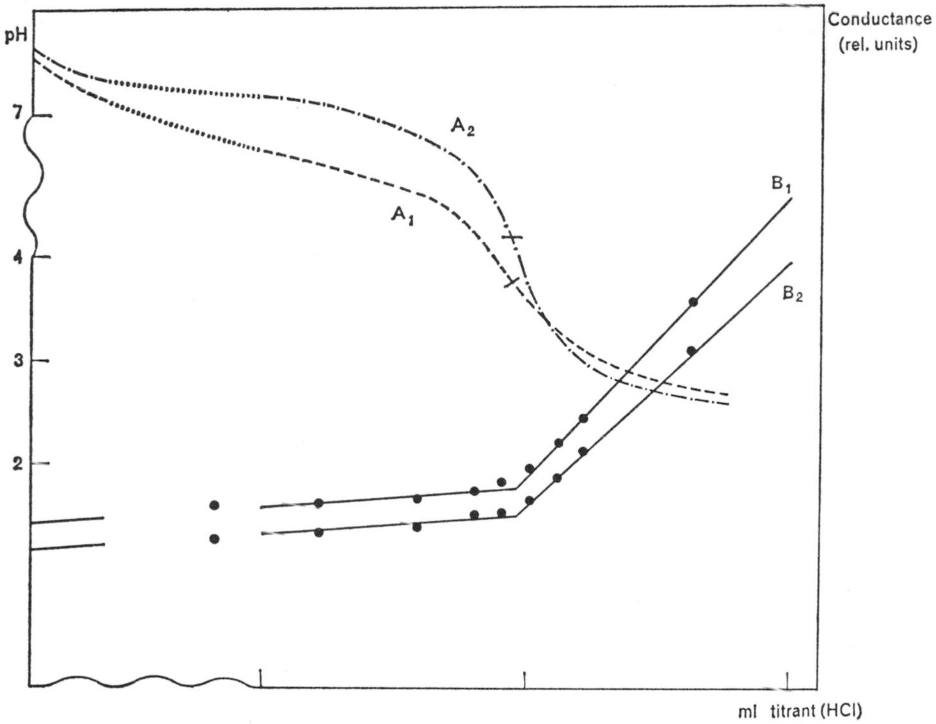

Figure 9.3. Titration curve of 0·001 M $NaHCO_3$ with HCl.

A_1. Potentiometric titration. A_2. Potentiometric titration under N_2 stream.
B_1. Conductometric titration. B_2. Conductometric titration under N_2 stream.

Two types of alkalinity can be determined, which correspond to the two inflexion points of a titration curve (see Fig. 9.4); Phenolphthalein Alkalinity (PA) and Total Alkalinity (TA).

The equivalence pH in a PA determination is equal to the pH of a $NaHCO_3$ solution of identical total ionic strength, temperature and concentration of total CO_2. For concentrations greater than about 0·2 mM of bicarbonate this pH is very near to 8·3, corresponding to the lower end of the range of phenolphthalein indicator.

The equivalence pH in a TA determination is equal to the pH of a CO_2 solution of identical ionic strength, temperature and concentration of total CO_2. This equivalence pH varies with the concentration of CO_2 present at the equivalence point, but it is usually within the pH values 4·2–5·4 (see Table 3.6). At the equivalence point the situation is described by the following equations.

$$[CO_2]_T = [\text{free } CO_2 + H_2CO_3] + [HCO_3^-] + [CO_3^{2-}] \qquad (9.15)$$
$$[H^+] = [HCO_3^-] + [OH^-] + 2\,[CO_3^{2-}]$$

and

$$K_1 = \frac{[H^+][HCO_3^-]}{[CO_2 + H_2CO_3]}$$

For alkalinities > 0·05 mM and conductivities < 400 μS cm^{-1} the simplified equation (9.16) may be used for calculation of the equivalence pH with a calculated error of 0·01 pH unit:

$$pH = -\tfrac{1}{2}\log([CO_2]_T K_1) = \tfrac{1}{2}(pK_1 - \log[CO_2]_T) \quad (9.16)$$

and for alkalinities > 0·2 mM equation (9.17) is sufficiently accurate:

$$pH = -\tfrac{1}{2}\log(K_1[CO_2]_T) \quad (9.17)$$

These formulae are derived from the formula for the endpoint of the alkalinity titration (Stumm and Morgan, 1970, with corrections):

$$(H^+)^4 + (H^+)^3K_1 + (H^+)^2(-CK_1 + K_1K_2 - K_w) - (H^+)K_1(CK_2 + K_w) - K_1K_2K_w = 0 \quad (9.18)$$

(9.16) is derived from this formula by assuming CO_3^{2-} is negligible and omitting the terms H^+K and K_1K_2 relative to $(H^+)^2$, while (9.17) is derived by neglecting (H^+) and (OH^-) relative to HCO_3^- and dropping all terms involving K_w and realising that $K_2 \ll K_1$ and $K_1^2 \ll 4\,K_1C$.

Equivalence pH calculated by A. Rebsdorf according to equation (9.16) are given in Table 3.6 for different values of $[CO_2]_T$ at 20°C, where $K_1 = 10^{-6\cdot 38}$. The values are 0·3–0·4 units lower than those given in Standard Methods (1965), but agree with those of Weber and Stumm (1963, page 1569).

A loss of less than 10% of the total CO_2 will only make the end point 0·02 pH units higher than that given in Table 3.6. It is fairly easy to keep losses to less than 10% in a routine titration if magnetic stirring is used.

9.8 DETERMINATION OF TOTAL CO_2

(*a*) By titration

For the determination of total CO_2 (used for primary productivity measurements by the ^{14}C method) a titration method and a calculation method may be used in waters where there is sufficient evidence that the alkalinity is caused mainly by CO_3^{2-}, HCO_3^- and OH^-.

The titration method is based on the principle that the difference in volume of titrant between the PA end point and the TA end point is equivalent to the total CO_2. The pH of the sample is first brought to 8·3 by careful addition of CO_2-free strong acid or base. Then the solution is titrated with strong acid to the TA end point as in a normal TA determination (Fig. 9.4).

(*b*) By calculation

From a knowledge of the initial pH and the total alkalinity it is possible to calculate the concentration of total CO_2:

$$[CO_2]_T = [CA]\left\{\frac{1 + [H^+]/K_1 + K_2/[H^+]}{1 + (2\,K_2/[H^+])}\right\} \quad (9.19)$$

It is convenient to represent the expression within braces by δ

$$[CO_2]_T = [CA]\delta = ([TA] - [OH^-])\delta$$

where $[CA]$ = carbonate alkalinity [mol l^{-1}]
$[CO_2]_T$ = total CO_2 concentration [mol l^{-1}]

Because K_1 and K_2 are temperature dependent so is δ. The factor δ is plotted in Fig. 9.5 as a function of temperature and pH. With programmable calculators it is easy to write one's own programme.

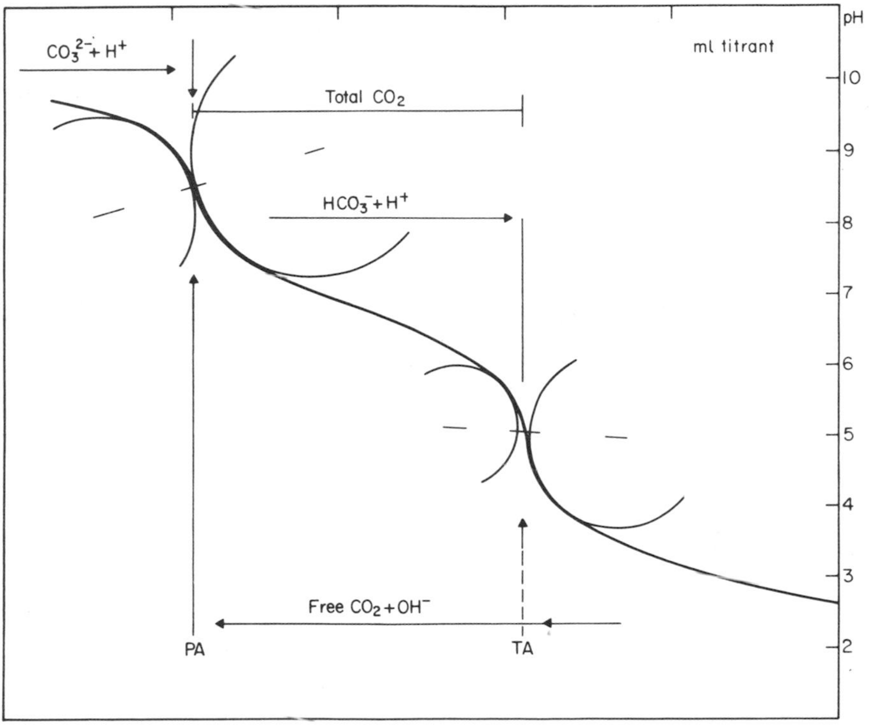

Figure 9.4. Titration curve of Na_2CO_3 with HCl or of H_2CO_3 with NaOH.
The following titrations are shown: $CO_3^{2-} + H^+ \rightarrow HCO_3^-$
$HCO_3^- + H^+ \rightarrow H_2O + CO_2$
free $CO_2 + OH^- \rightarrow HCO_3^-$

For the calculations the reactions described by (9.11_1) and (9.11_2) are considered. Equations (9.2) and (9.15) are used together with (9.20).

$$[TA] = [HCO_3^-] + 2[CO_3^{2-}] + [OH^-] - [H^+] \quad (9.20)$$

where $[TA]$ = total alkalinity.

For evaluating the influence of the ionic strength on δ the correction values to be added to δ have been calculated for different ionic strengths and pH values. Carbonate alkalinity [CA] is also influenced by the ionic strength at higher pH values. Both corrections are incorporated in Table 3.5.

The calculations are based on 'Standard composition water' (Rodhe, 1949; Karlgren, 1962), which is the average composition of fresh water, calculated for the whole world. The composition is indicated in Table 9.5.

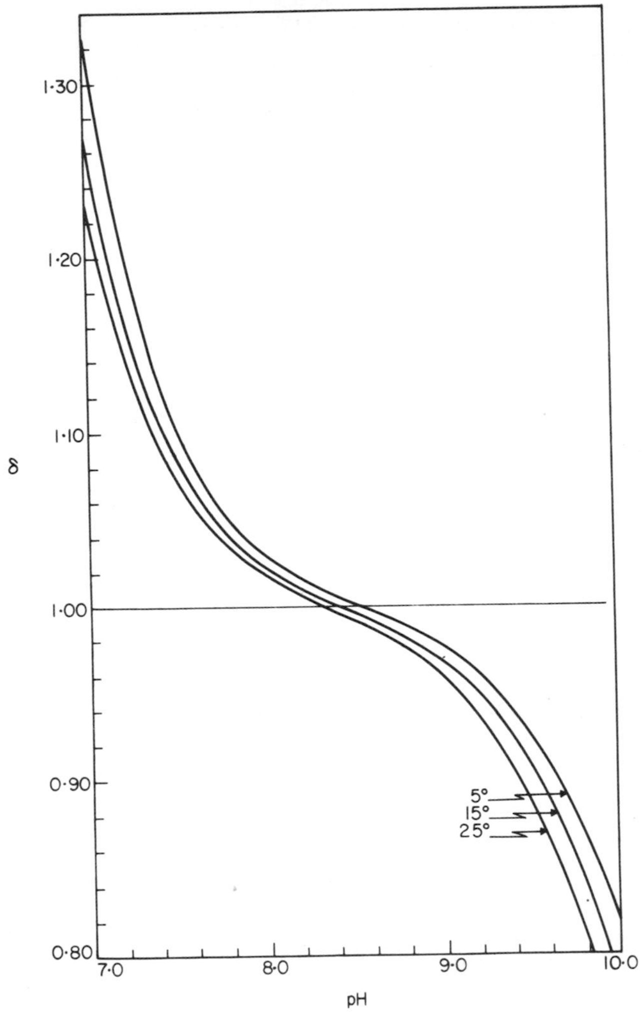

Figure 9.5. The factor δ (see text) from Vollenweider (1971).
The factor is pH and temperature dependent, and to a lesser extent ionic strength dependent. This graph is strictly valid for zero ionic strength. Table 3.5 shows the magnitude of error to be expected at 20°C.

Table 9.5. 'Standard composition water'. Data from L. Karlgren. Vattenkemiska Analysmetoder, Limnologiska Institutionen, Uppsala, 1962. The figures are % of total anion concentration.

Ca	Mg	Na	K	HCO_3	SO_4	Cl
63·5	17·4	15·7	3·4	74·3	15·7	10·0

9.9 POTENTIAL DIFFERENCES

When a rod of a metal (other than a noble metal) is placed in a solution of a salt of the same metal a potential difference will appear between metal and solution.

Metal ions tend to go into solution. As the metal ions are positive, the rod becomes electrically negative relative to the solution. This electric potential difference tends to hold back the positive metal ions, which are, moreover, already present in the solution. The rod will almost instantaneously reach equilibrium with the solution when as many metal ions are dissolving as are being deposited per unit of time.

According to Nernst the potential difference E (in volts) depends on the tendency of the metal to dissolve and on the concentration of metal ions in solution, C, as given in equation (9.21). When this concentration exceeds a certain value ('saturation value') the process is reversed.

$$E = \frac{RT}{nF} \ln \frac{K}{C} \qquad (9.21)$$

where R is the molar gas constant, T is the absolute temperature, n is the number of electrons involved, F is the faraday, K is a constant.

A metallic-rod in a solution of its salt is called an *indicator electrode*. It is not possible to measure the potential of one electrode. Two rods must therefore be placed in two solutions with concentrations C_1 and C_2 and connected by a salt bridge. The potential difference between the two rods will be

$$E = E_1 - E_2 = \frac{RT}{nF} \ln \frac{C_2}{C_1} \qquad (9.22)$$

There may also be a potential difference between the two solutions due to a difference in diffusion rate of anions and cations, a so-called diffusion potential=junction potential. By using an appropriate salt bridge between the two solutions the size of this potential can be made insignificant.

The KCl salt bridge is used for this purpose with the calomel electrode.

The potential of an electrode is measured by comparing it with the potential of an arbitrary standard electrode. To obtain this standard potential the metallic rod is placed in a saturated solution of one of its salts. The most commonly used electrode consists of Hg as the metal in a slurry of Hg_2Cl_2 (calomel-electrode) or Hg_2SO_4 in contact with 1 M or saturated KCl (or K_2SO_4).

This type of electrode is called a *reference electrode*. Equation (9.21) holds also for H_2-gas in contact with H^+. The H_2-gas is kept in equilibrium in the solution by bubbling H_2 at one atmosphere pressure over spongy platinum (from which the electrical connection is taken). The potential of this electrode depends only on the H^+ concentration of the solution (and the temperature).

When this electrode is placed in a solution containing 1 M H^+ equation (9.21) becomes

$$E = \frac{RT}{nF} \ln K$$

Measuring this potential relative to a calomel electrode gives

$$E_0 = \frac{RT}{nF} \ln K - E_r \qquad (9.23)$$

where E_r is the potential of the calomel electrode. In a second solution, with $[H^+] = C_H$ the potential becomes

$$E_c = \frac{RT}{nF} \ln \frac{K}{C_H} - E_r \qquad (9.24)$$

Subtracting (9.23) from (9.24) gives

$$E_c = E_0 - \frac{RT}{nF} \ln C_H \tag{9.25}$$

At 18°C $(RT/nF) = 0{\cdot}0577$ (in which the conversion from ln to $\log_{10}$ is included). Then equation (9.25) becomes

$$-\log C_H = \text{pH} = \frac{E_c - E_0}{0{\cdot}0577}$$

Thus when $E_c - E_0$ equals 57·7 mV the pH equals 1, when $E_c - E_0 = 2 \times 57{\cdot}7$ mV, the pH = 2. This hydrogen electrode can therefore be used as an *indicator electrode* for H^+ concentrations, with the same electrode in 1 M $H^+ + 1$ atm H_2 as reference electrode. The combination of the H_2-electrode and calomel electrode can be used for a direct determination of the pH, after calibration in a buffer with a known pH.

It is more convenient to use a simpler H^+ sensitive electrode, the glass electrode. This is a small thin glass bulb (ca. 0·05 mm thick) which is filled with an electrolyte. The potential of this electrode depends on $[H^+]$ in the same way as the normal H_2-electrode. It must be calibrated against a buffer solution with a known pH value.

The pH scale was originally defined by the pH of a 0·1 M potassium hydrogen-phthalate solution at 20°C, and this solution is very useful for calibrating glass electrodes too. The pH of this solution is only slightly temperature dependent. Nowadays the operational pH scale is defined by reference to five buffer solutions similar to those of § 3.2.4 (McGlashan, 1971).

The actual numerical value of the potential can be defined in two ways. Either the normal H_2-electrode or the calomel electrode is arbitrarily set equal to zero. The first system is at present more commonly used.

9.10 OXIDATION-REDUCTION POTENTIALS

When a noble metal such as Pt is placed in a solution of, for example, Fe^{3+} ions, the Pt becomes positively charged. The Fe^{3+} ions react with the electrons in the Pt-metal according to

$$Fe^{3+} + e^- \rightleftharpoons Fe^{2+} \tag{9.26}$$

When the Pt-rod is placed in a solution of Fe^{2+} ions the rod will become negatively charged.

The same processes occur when the Pt-rod is placed in solutions of salts which take up or give off electrons, for example Cr^{3+}, $Fe(CN)_6^{4-}$, MnO_4^-, $Cr_2O_7^{2-}$.

These potential differences are called oxidation-reduction potentials (= redox potentials). They depend on the nature and the concentration of the ions involved and are therefore fundamentally not different from the potentials described in § 9.9.

When the Pt-rod is placed in a solution containing both Fe^{2+} and Fe^{3+} ions the Pt will become positive or negative, depending on whether ferric or ferrous ions are present in excess. At one ratio of Fe^{3+} to Fe^{2+} the Pt-rod remains neutral. The reaction velocities in reaction (9.26) to the left or to the right are then equal, so that the equilibrium constant

$$K = \frac{[Fe^{2+}]}{[Fe^{3+}]} \tag{9.27}$$

The Pt-rod therefore remains neutral in a solution in which the ratio of Fe^{2+} to Fe^{3+} equals K.

The E.M.F. between two Pt-rods placed in two separate solutions (A and B) in which the concentrations of the oxidized ions are $[Ox_A]$ and $[Ox_B]$ respectively and the concentrations of the reduced ions $[Red_A]$ and $[Red_B]$ (and which solutions are in electrical contact with each other) is

$$E = \frac{RT}{nF} \ln \frac{[Ox_A]}{[Red_A]} - \frac{RT}{nF} \ln \frac{[Ox_B]}{[Red_B]} \tag{9.28}$$

provided that the oxidation-reduction reaction is reversible. For reaction (9.26), equation (9.28) becomes ($n = 1$)

$$E = \frac{RT}{F} \ln \frac{[Fe_A^{3+}]}{[Fe_A^{2+}]} + \frac{RT}{F} \ln \frac{[Fe_B^{2+}]}{[Fe_B^{3+}]} \tag{9.29}$$

When solution B is in equilibrium equation (9.29) becomes:

$$E = \frac{RT}{F} \ln \frac{[Fe_A^{3+}]}{[Fe_A^{2+}]} + \frac{RT}{F} \ln K$$

or in general, replacing $RT/F \ln K$ by E_0

$$E = E_0 + \frac{RT}{F} \ln \frac{[Ox]}{[Red]} \tag{9.30}$$

in which [Ox] means the concentration of the oxidized ions and [Red] that of the reduced ions. E_0 is the potential of the Pt electrode in a solution in which [Ox] = [Red], measured against for instance a calomel electrode. The lower the normal oxidation reduction potential, the stronger the reducing capacity of the substance involved. It is customary to set the potential of the normal H_2-electrode equal to zero. A solution having a negative oxidation-reduction potential is called reducing (giving off electrons), a solution with a positive potential is oxidizing (taking up electrons).

9.11 pH DEPENDENT REDOX POTENTIALS

The redox potentials of some organic compounds are sensitive to the pH of the solution. This can be demonstrated for instance with the reaction

$$\underset{\text{quinone}}{C_6H_4O_2} + 2H^+ + 2e^- \rightleftharpoons \underset{\text{hydroquinone}}{C_6H_6O_2}$$

If [quinone] = [hydroquinone] equation (9.30) becomes

$$E = E_0 + \frac{RT}{F} \ln [H^+]$$

or at 18°C

$$E = 0{\cdot}704 + 0{\cdot}0577 \log [H^+] \tag{9.31}$$

By using quinhydrone—a slightly soluble compound formed by the combination of one molecule of quinone and one molecule of benzohydroquinone—either the oxidation reduction potential or the hydrogen ion concentration can be measured. Quinhydrone is however usually used only to calibrate Pt-electrodes, because for routine pH determinations the glass electrode is more convenient. When calibrating Pt electrodes quinhydrone is added to a buffer solution of known pH and the potential against the H_2-electrode is calculated with equation (9.31):

$$E = 0{\cdot}704 - 0{\cdot}0577\ \text{pH}$$

If a saturated calomel reference electrode is used the calomel electrode will be negative with respect to the quinhydrone electrode and $E = 0{\cdot}45 - 0{\cdot}058$ pH.

As hydroquinone is a weak acid ($K \sim 10^{-10}$) the dissociation becomes appreciable above pH $= 8{\cdot}5$ and the hydroquinone electrode should not then be used.

For purposes of comparison the potentials of pH sensitive systems are calculated to E_7, that is the potential the system would have at pH $= 7$. The calculation is carried out by assuming a rise of one pH unit to be equivalent to a decrease of 58 mV.

In natural waters and muds the 'apparent' potential difference is usually between $-0{\cdot}1$ V (O_2 free) and $+0{\cdot}5$ V (O_2 saturated) relative to the calomel reference electrode. In theory water saturated with O_2 should have a value of about 0·8 V. The reason for the discrepancy between 0·5 V and 0·8 V probably lies in the fact that the Pt-electrode becomes covered with a film of oxide and is no longer completely reversible.

The negative values are normally reached only in sediments. Waters with a potential lower than 0·1 to 0·2V are generally called 'reducing', although this is a relative description, as it depends also on the system to be reduced.

Mortimer (1942) has given a list of approximate redox potential ranges, within which certain reductions proceeded actively (see Table 9.6).

Table 9.6. Reduction of some systems and prevalent redox potentials and O_2 concentrations, Mortimer (1942).

reduction reaction	redox potential* [mV]	prevalent O_2 concentration [mg l^{-1}]
NO_3^-: NO_2^-	450–400	4
NO_2^-: NH_3	400–350	0·4
Fe^{3+}: Fe^{2+}	300–200	0·1
SO_4^{2-}: S^{2-}	100– 60	0·0

*The lower potential is the limit below which Mortimer could not detect the oxidized phase.

These 'apparent' potential differences should not be confused with the thermodynamic potential differences, which are based on reversible systems. The 'apparent' potential differences are caused by a mixture of inorganic and organic compounds most of which do not take direct part in oxidation-reduction reactions. This measured difference has a purely empirical significance in describing the oxidation state of the system.

It is important to realise that the decrease of redox potentials in sediments mentioned in Table 9.6 is caused by the depletion of O_2 and the release of organic compounds by the mud. The reactions follow the redox potential; the potentials mentioned are not the potentials of the systems in column 1. For example the NO_2^- in the first line is formed by the oxidation of NH_3, which is produced continuously by the sediments even while O_2 is still measurable. The NO_3^- is reduced to N_2. When the oxygen concentration falls further NH_3 is no longer oxidised and NO_2^- disappears (Golterman, 1975).

9.12 POTENTIOMETRIC TITRATIONS

All potential differences between reference electrodes and indicator electrodes (such as H^+ indicator or Ag^+ indicator and oxidation reduction electrodes) can be used to follow a titration in order to determine the end point.

When the potential is E Volt the addition of a fixed quantity of for instance a Ag^+ solution (V ml) in the case of a Cl^- determination causes a change: ΔE. It can be shown that the $\Delta E/\Delta V$ is maximal and therefore $\Delta^2 E/\Delta V^2 = 0$ at the equivalence point. A graph of E plotted versus volume of reagent added therefore shows a rapid change near the equivalence point, which has a potential which is characteristic of the reaction. It is, however, more accurate to determine the end point by plotting the rate of change of E against V ($\Delta E/\Delta V$ vs. V) than by using the E of the end point itself, as a small error in E may introduce a larger eror in the quantity of reagent necessary. When the rate of change of E is used, the actual value of E itself need not be known.

Potentiometric titrations can be used in limnology for the determination of alkalinity, chloride, and for the oxidative or reductive capacity of a water (i.e. oxygen, oxidability, ferrous compounds). Potentiometric titrations can be made specific and can be automated easily.

9.13 DEAD-STOP END POINT TITRATIONS

The detailed mechanism of the dead-stop titration is complicated. It is explained in standard works (for example Lingane, 1958). There follows a very brief outline of the essentials.

The dead-stop end point titration is an amperometric or potentiometric titration with two identical Pt-electrodes. One of these is made the indicator electrode by electrical polarization with a small constant voltage (10–500 mV); the other one serves as a reference electrode.

During the titration the electric current or the voltage is measured. At the end point there is a sharp change in current or voltage.

The configuration of the titration curve depends on whether the titrate, the titrant or both are electroactive, i.e. reducible. When the titrate is not reducible the current remains constant (practically zero) before, but increases beyond the end point. When the titrant is not reducible there is first a decrease of current, which becomes constant after the end point. When both titrant and titrate are reducible a V-shaped trace is produced.

A different type of amperometric titration curve will be obtained when the electrodes exchange the roles of 'indicator' and 'reference' during the titration (when titrate and titrant both establish reversible redox couples).

With choice of different electrodes ('stirring' or 'dropping') and changes in the voltage numerous applications are possible. In limnochemistry the main applications are the titrations involving I_2, $Cr_2O_7^{2-}$, Ce^{4+}, MnO_4^-, CrO_4^{2-}, SO_4^{2-} (with Ba^{2+}).

Dead-stop titrators may be bought, but they can be made relatively easily. A circuit diagram is shown as Fig. 9.6. The final adjustment of resistors depends on the electrodes being used. As only low voltages and small currents are necessary the method is suitable for field conditions. The use of a micro burette is advisable.

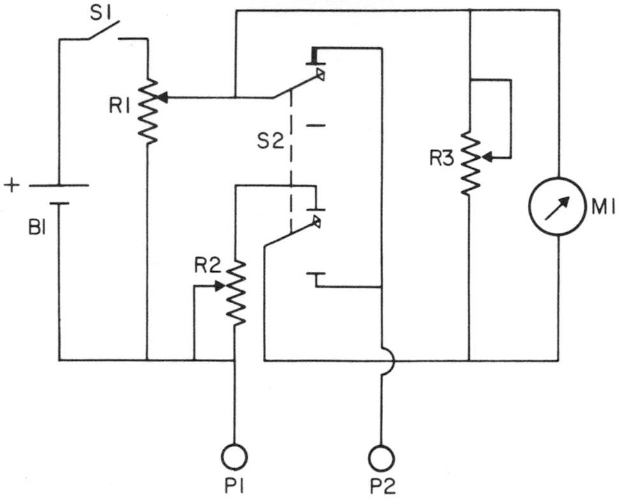

Figure 9.6. Circuit for dead-stop end point titrator, according to van Asdonk. B1: Mallory 'Duracell' 1·35 V; S1: SPST switch; S2: DPDT switch; R1: Wirewound potentiometer 500 Ω; R2: Preset potentiometer 10 kΩ set to about 8·5 kΩ; R3: Preset potentiometer 10 kΩ set to about 9·5 kΩ; M1: Tautband meter, 7·5 mA fsd; P1, P2: electrodes Pt gauze, 48 mesh, 3 × 3 cm.

Accuracy

In macro titrations (e.g. 0·1 M titrant, even when a micro burette is used) it is often possible to obtain an error as small as 0·1%, even without graphing the current readings against volume of titrant. In the case of a micro titration it is necessary to graph current against volume of titrant as the trace is always curved at the end point. A linear extrapolation is possible using points sufficiently distant from either side of the end point. The accuracy in this case is no better than with a potentiometric titration, but the apparatus is much simpler.

REFERENCES

A prefatory asterisk (*) indicates a handbook or work of wide scope

AFGHAN B.K., GOULDEN P.D. and RYAN J.F. 1972 Automated fluorometric method for determination of boron in waters, detergents and sewage effluents. *Wat. Res.* **6** 1475–1485.

AFGHAN B.K. and RYAN J.F. 1975 A modified procedure for the determination of nitrate in sediments and some natural waters. *Envir. Letters* **9** 59–73.

AGEMIAN H. and CHAU A.S.Y. 1975 An atomic absorption method for the determination of 20 elements in lake sediments after acid digestion. *Analytica chim. Acta* **80** 61–66.

*ALLEN S.E., GRIMSHAW H.M., PARKINSON J.A. and QUARMBY C. 1974 *Chemical Analysis of Ecological Materials.* Blackwell, Oxford, 565 pp.

ALSTERBERG G. 1926 Die Winklersche Bestimmungsmethode für in Wasser gelösten, elementaren Sauerstoff sowie ihre Anwendung bei Anwesenheit oxydierbarer Substanzen. *Biochem. Z.* **170** 30–75.

ANDERSON E.R. 1952 Energy-budget studies: Water loss investigations, Vol. 1 Lake Hefner Studies. *Tech. Rept. Circ. U.S. Geol. Surv.* **229** 71–119.

ARMOUR J.A. and BURKE J.A. 1970 Method for separating polychlorinated biphenyls from DDT and its analogs. *Jnl. Ass. Off. Analyt. Chem.* **53** 761–768

ARMSTRONG F.A.J. and TIBBITS S. 1968 Photochemical combustion of organic matter in seawater for nitrogen, phosphorus and carbon determination. *J. mar. biol. Ass. U.K.* **48** 143–152.

ARMSTRONG F.A.J., WILLIAMS P.M. and STRICKLAND J.D.H. 1966 Photo-oxidation of organic matter in sea water by ultra-violet radiation; analytical and other applications. *Nature, Lond.* **211** 481–483.

ARNOLD G.P. 1975a The measurement of irradiance with particular reference to marine biology. In Evans, Rackham and Bainbridge (Ed.) *q.v.* 1–25.

ARNOLD G.P. 1975b Standards of spectral power distribution for measuring spectral irradiance. In Evans, Rackham and Bainbridge (Ed.) *q.v.* 573–583.

ATKINS W.R.G. and POOLE H.H. 1929 Methods for the photo-electric and photo-chemical measurement of daylight. *Biol. Rev.* **5** 91–113.

ÅBERG B. and RODHE W. 1942 Uber die Milieufaktoren in einigen sudschwedischen Seen. *Symb. bot. upsal.* **5** 1–256.

BACHMANN R.W. and GOLDMAN C.R. 1964 The determination of microgram quantities of molybdenum in natural waters. *Limnol. Oceanogr.* **9** 143–146.

BAKER C.D., BARTLETT P.D., FARR I.S. and WILLIAMS G.I. 1974 Improved methods for the measurement of dissolved and particulate organic carbon in fresh water and their application to chalk streams. *Freshwater Biol.* **4** 467–481.

*BATES R.G. 1964 *Determination of pH, Theory and Practice.* Wiley, New York, London. 435 pp.

BAUER L. 1952 Trennung der Karotinen und Chlorophylle mit hilfe der Papierchromatographie. *Naturwissenschaften* **39** 88.

BEADLE L.C. 1958 Measurement of dissolved oxygen in swamp waters. Further modification of the Winkler method. *J. exp. Biol.* **35** 556–566.

BEDFORD J.W. 1974 The use of polyurethane foam plugs for extraction of polychlorinated biphenyls (PCB's) from natural waters. *Bull. Environ. Contam. & Toxicol.* **12** 622–625.

BILIKOVA A. 1973 Determination of cobalt in water by using β nitroso-α-naphthol. *Chem. Abstr.* **78** 140246v.

BLAŽKA P. 1966a Bestimmung der Proteine in Material aus Binnengewässern. *Limnologica* **4** 387–396.

BLAŽKA P. 1966b Bestimmung der Kohlenhydrate und Lipide. *Limnologica* **4** 403–418.

*BOLLINGER H.R., BREUNER M., GANSHIRT H., MANGOLD H.K., SEIDLER H., STAHL W.D. and E. 1965 in *Thin-Layer Chromotography; a Laboratory Handbook.* E. Stahl (Ed.) Academic Press, New York.

*BRADSTREET R.B. 1965 *The Kjeldahl Method for Organic Nitrogen.* Academic Press, New York, London. 239 pp.

BREIDENBACH A.W., LICHTENBERG J.J., HENKE C.F., SMITH D.J., EICHELBERGER J.W. and STIERLI H. 1964 The identification and measurement of chlorinated hydrocarbon pesticides in surface water. *U.S. Publ. Hlth. Serv. Publs. Wash. nr.* 1241.

BRIGGS R. and VINEY M. 1964 The design and performance of temperature compensated electrodes for oxygen measurements. *J. scient. Instrum.* **41** 78–83.

BROWN S.R. 1968 Absorption coefficients of chlorophyll derivatives. *J. Fish. Res. Bd. Can.* **25** 523–540.

BRYAN J.R., RILEY J.P. and WILLIAMS P.J. 1976 A Winkler procedure for making precise measurements of oxygen concentration for productivity and related studies. *J. Exp. Mar. Biol. & Ecol.* **21** 191–197.

BUCHOLZ C.F. 1805 Mémoire sur l'Urane. *Annls. Chim. Phys.* **56** 142–151.

BURNHAM A.K., CALDER G.V., FRITZ J.S., JUNK G.A., SVEC H.J. and WILLIS R. 1972 Identification and estimation of neutral organic contaminants in potable water. *Analyt. Chem.* **44** 139–142.

BURNS E.R. and MARSHALL C. 1965 Correction for chloride interference in the chemical oxygen demand test. *J. Wat. Pollut. Control. Fed.* **37** 1716–1721.

BUSCH A.W. 1966 Energy, total carbon, and oxygen demand. *Wat. Resour. Res.* **2** 59–69.

BUSCH A.W. 1967 Total carbon analysis in water pollution control. In Golterman and Clymo (Ed.) *q.v.* 133–143.

CARLUCCI A.F. and SILBERNAGEL S.B. 1966a Bioassay of Sea Water (1). A ^{14}C-uptake method for the determination of vitamin B_{12} in sea water. *Can. J. Microbiol.* **12** 175–183.

CARLUCCI A.F. and SILBERNAGEL S.B. 1966b Bioassay of Sea Water (2). Methods for the determination of concentrations of dissolved vitamin B_{12} in sea water. *Can. J. Microbiol.* **12** 1079–1089.

CARLUCCI A.F. and SILBERNAGEL S.B. 1967a Bioassay of Sea Water (4). The determination of dissolved biotin in sea water using ^{14}C-uptake by cells of *Amphidinium carteri. Can. J. Microbiol.* **13** 979–986.

CARLUCCI A.F. and SILBERNAGEL S.B. 1967b Determination of vitamins in sea water. In Golterman and Clymo (Ed.) *q.v.* 239–244.

CARPENTER J.H. 1965 The accuracy of the Winkler method for dissolved oxygen analysis. *Limnol. Oceanogr.* **10** 135–140.

CARPENTER J.H. 1966 New measurements of oxygen solubility in pure and natural water. *Limnol. Oceanogr.* **11** 264–277.

CHABROL E. and CHARONNAT R. 1937 Une nouvelle réaction pour l'étude des lipides. *Presse Med.* **96** 1713.

CHABROL E., BÖZÖRMÉNYI M. and FALLOP P. 1949 Les lipides du sérum sanguin sous l'angle de la réaction sulpho-phosphovanillique. *Sem. Hôp. Paris* **25** 3446.

CHAN K.M. and RILEY J.P. 1966a The determination of vanadium in sea water and natural waters, biological materials and silicate sediments and rocks *Analytica. chim. Acta* **34** 337–345.

CHAN K.M. and RILEY J.P. 1966b The determination of molybdenum in natural waters, silicates and biological materials. *Analytica. chim. Acta* **36** 220–229.

CHENG K.L., MELSTED S.N. and BRAY R.H. 1953 Removing interfering metals in the versenate determination of calcium and magnesium. *Soil. Sci.* **75** 37–40.

*CHRISTIAN G.D. and FELDMAN F.J. 1970 *Atomic Absorption Spectroscopy:* Applications in Agriculture, Biology and Medicine. Wiley-Interscience. New York. 490 pp.

*CHRISTIE W.W. 1973 *Lipid Analysis*, Isolation, Separation, Identification and Structural Analysis of Lipids. Pergamon Press, Oxford. 338 pp.

COBURN J.A., VALDMANIS I.A. and CHAU A.S.Y. 1977 Evaluation of XAD-2 for multiresidue extraction of organochlorine pesticides and polychlorinated biphenyls from natural waters. *Jnl. Ass. Off. Analyt. Chem.* **60** 224.

COCHRANE W.P. and CHAU A.S.Y. 1971 Chemical derivatization techniques for confirmation of organochlorine residue identity. *Advances in Chemistry Series* **104** 11–26.

COOKE A.S. 1972 The effects of DDT, Dieldrin and 2,4-D on amphibian spawn and tadpoles. *Environ. Poll.* **3** 51–68.

DARWIN C. 1859 *The Origin of Species.* John Murray, London.

DAVISON W. 1976 Comparison of differential pulse and D.C. sampled polarography for the determination of ferrous and manganous ions in lake water. *J. electroanal. Chem.* **72** 229–237.

DAVISON W. 1978 Defining the electroanalytically measured species in a natural water sample. *J. electroanal. Chem.* **87** 395–404.

DENNY P. 1972 Lakes of South-Western Uganda 1. Physical and chemical studies of Lake Bunyonyi. *Freshwater Biol.* **2** 143–158.

DENNY P. 1977 A simple field technique for volumetric titrations. *Freshwater Biol.* **7** 185–186.

DINNIN J.I. 1960 Releasing effects in flame photometry. Determination of calcium. *Analyt.Chem.* **32** 1475.

DORE W.G. 1958 A simple chemical lightmeter. *Ecology* **39** 151–153.

DUSSART B. and FRANCIS-BOEUF C. 1949 Technique du dosage de l'oxygène dissous dans l'eau basé sur la méthode de Winkler. *Circul. Cent. Rech. Étud. océanoger. Instr. Techn nr.* 1.

DUURSMA E.K. 1961 Dissolved organic carbon, nitrogen and phosphorus in the sea. *Neth. J. Sea Res.* **1** 1–148.

DUURSMA E.K. 1974 The fluorescence of dissolved organic matter in the sea. In *Optical Aspects of Oceanography.* N.G. Jerlov and E. Steemann Nielsen (Ed.) Academic Press, London. pp. 237–256.

DUURSMA E.K. and ROMMETS J.W. 1961 Interprétation mathématique de la fluorescence des eaux douces, saumâtres et marines. *Neth. J. Sea Res.* **1** 391–405.

*E.A.W.A.G. 1973 *Tabellen der Sättigungswerte von Wasser.*

EDMOND J.M. 1970 High precision determination of titration alkalinity and total carbon dioxide content of sea water by potentiometric titration. *Deep Sea Res.* **17** 737–750.

EDWARDS D.P. and EVANS G.C. 1975 Problems involved in the design of apparatus for measuring the spectral composition of daylight in the field. In Evans, Rackham and Bainbridge (Ed.) *q.v.* 161–187.

EFFENBERGER M. 1962 Konduktometrische Bestimmung des organischen Kohlenstoffs in Gewässern. *Sb. vys. šk. chem.-technol. Praze* **6** 471–493.

EFFENBERGER M. 1967 A simple flow-cell for the continuous determination of oxidation reduction potential. In Golterman and Clymo (Ed.) *q.v.* 123–126.

*ELWELL W.T. and GIDLEY J.A.F. 1966 *Atomic Absorption Spectroscopy.* Pergamon Press, London. 138 pp.

EVANS G.C. 1969 The spectral composition of light in the field. 1. Its measurement and ecological importance. *J. Ecol.* **57** 109–125.

*EVANS G.C., RACKHAM O. and BAINBRIDGE R. (Ed.). 1975 *Light as an Ecological Factor II.* 16th Symposium of the British Ecological Society. Blackwell, Oxford. 616 pp.

*EVERETT K. 1967 Handling Perchloric Acid and Perchlorates. In *Handbook of Laboratory Safety.* N.V. Steere (Ed). Chemical Rubber Publishing Co., Cleveland, Ohio. 568 pp.

FABRICAND B.P., SAWYER R.R., UNGAR S.G. and ADLER S. 1962 Trace metal concentrations in the ocean by atomic absorption spectroscopy. *Geochim. cosmochim. Acta* **26** 1023–1027.

FALES F.W. 1951 The assimilation and degradation of carbohydrates by yeast cells. *J. biol. Chem.* **193** 113–124.

FARKAS T. and HERODEK S. 1964 The effect of environmental temperature on the fatty acid composition of Crustacean plankton. *J. Lipid Res.* **5** 369–373

FASSEL V.A. and MOSOTTI V.G. 1965 Evaluation of spectral continua as primary light sources in atomic-absorption spectroscopy. *Coll. Spectr. Internat.* 12.

FAY P.F. 1970 Photostimulation of N_2 fixation. *Biochim. biophys. Acta* **216** 353–356.

FEE E.J. 1976 The vertical and seasonal distribution of chlorophyll in lakes of the Experimental Lakes Area, northwestern Ontario: implications for primary production estimates. *Limnol. Oceanogr.* **21** 767–783.

FISHMAN M.J. and DOWNS S.C. 1966 Methods for analysis of selected metals in water by atomic absorption. U.S.Geol.Surv. Water Supply Paper. 1540–C 23–45.

FISHMAN M.J. and ERDMANN D.E. 1975 Water analysis. *Analyt. Chem.* **47** 334 R–361 R.

FISHMAN M.J. and SKOUGSTAD W.M. 1964 Catalytic determination of vanadium in water. *Analyt. Chem.* **36** 1643–1646.

FISHMAN M.J. and SKOUGSTAD W.M. 1965 Rapid field and laboratory determination of phosphate in natural water. U.S. Geol. Surv. Prof. Paper 525–B (*Anal. Techn.*) B 167–B 169.

*FLASCHKA H.A. 1964 *E.D.T.A. Titrations;* an introduction to theory and practice. 2nd ed. Pergamon Press, Oxford. 144 pp.

FOX H.M. and WINGFIELD C.A. 1938 A portable apparatus for the determination of oxygen dissolved in a small volume of water. *J. exp. Biol.* **15** 437–443.

FØYN E. 1955 The oxymeter. *Fisk Dir. Skr. Havundersøk.* **11** 1.
FØYN E. 1967 Density and oxygen recording in water. In Golterman and Clymo (Ed.) *q.v.* 127–132.

GEORGE D.G. 1976 A pumping system for collecting horizontal plankton samples and recording continuously sampling depth, water temperature, turbidity and *in vivo* chlorophyll. *Freshwater Biol.* **6** 413–419.
GLEBOVICH T.A. 1963 Determination of boron in waters with H-resorcinol. *Chem. Abstr.* **61** nr. 14359b.
GOERLITZ D.F. and BROWN E. 1972 Methods for analysis of organic substances in water. Techniques of water-resources investigations of the United States Geological Survey, Book 5, Chapter A3. United States Government Printing Office. Washington.
GOLTERMAN H.L. 1967a Opening address. In Golterman and Clymo (Ed.) *q.v.* 10–11.
GOLTERMAN H.L. 1967b Tetraethylsilicate as a 'molybdate unreactive' silicon source for diatom cultures. In Golterman and Clymo (Ed.) *q.v.* 56–62.
GOLTERMAN H.L. 1971 The determination of mineralization in correlation with the estimation of net primary production with the oxygen method and chemical inhibitors. *Freshwater Biol.* **1** 249–256.
*GOLTERMAN H.L. 1975 *Physiological Limnology*. Elsevier, New York, Amsterdam. 489 pp.
GOLTERMAN H.L. and CLYMO R.S. (Ed). 1967 'Chemical Environment in the Aquatic Habitat'. Proceedings of an IBP symposium 10–16 October, 1966. Royal Netherlands Academy of Sciences, Amsterdam. 322 pp.
GOLTERMAN H.L. and WÜRTZ I.M. 1961 A sensitive, rapid determination of inorganic phosphate in presence of labile phosphate esters. *Analytica chim. Acta* **25** 295–298.
GOMORI 1955 in *Methods in Enzymology*. S.P. Colowick and N.O. Kaplan (Ed.) Vol. 1 Academic Press, New York 835 pp.
GOODENKAUF A. and ERDEI J. 1964 Identification of chlorinated hydrocarbon pesticides in river water. *J. Am. Wat. Wks Ass.* **56** 600–606.
GOULDEN P.D. and AFGHAN B.K. 1970 An automated method for determining mercury in water. Inland Waters Branch, Department of Energy, Mines and Resources. *Techn. Bull.* 27.
GRAAF I.M. de and GOLTERMAN H.L. 1967 A rediscovered determination of chemical oxygen demand with chromium compounds. In Golterman and Clymo (Ed.) *q.v.* 166–168.
GRAN G. 1952 Determination of the equivalence point in potentiometric titrations. Part II. *Analyst, Lond.* **77** 661–671.
GRAVITZ N. and GLEYE L. 1975 A photochemical side reaction that interferes with the phenol-hypochlorite assay for ammonia. *Limnol. Oceangor.* **20** 1015–1017.
GREENHALGH R. and RILEY J.P. 1962 The development of a reproducible spectrophotometric curcumin method for determining boron, and its application to sea water. *Analyst, Lond.* **87** 970–976.

HALLDAL P. 1967 Ultraviolet action spectra in algology. *Photochem. & Photobiol.* **6** 445–460.
HANES C.S. 1929 An application of the method of Hagedorn and Jensen to the determination of larger quantities of reducing sugars. *Biochem. J.* **23** 99–106.
HANSEN A.L. and ROBINSON R.J. 1953 The determination of organic phosphorus in sea water with perchloric acid oxidation. *J. mar. Res.* **12** 31–42.
*HARNED H.S. and OWEN B.B. 1958 *The Physical Chemistry of Electrolytic Solutions*; 3rd ed. Reinhold, New York. 838 pp.
HART I.C. 1967 Nomograms to calculate dissolved-oxygen contents and exchange (mass-transfer) coefficients. *Wat. Res.* **1** 391–395.
HARWOOD J.E. and KÜHN A.L. 1970 A colorimetric method for ammonia in natural waters. *Wat. Res.* **4** 805–811.
HARWOOD J.E., van STEENDEREN R.A. and KÜHN A.L. 1969 A rapid method for orthophosphate analysis at high concentrations in water. *Wat. Res.* **3** 417–423.
HEANEY S.I. 1978 Some observations on the use of the *in vivo* fluorescence technique to determine chlorophyll a in natural populations and cultures of freshwater phytoplankton. *Freshwater Biol.* **8** (in press).
HENDEL Y. 1973 Replacement of platinum vessels with a pressure device for acid dissolution in the rapid analysis of glass by atomic-absorption spectroscopy. *Analyst, Lond.* **88** 450–451.
HENRIKSEN A. 1970 Determination of total nitrogen, phosphorus and iron in fresh water by photo-oxidation with ultraviolet radiation. *Analyst, Lond.* **95** 601–608.
HERRMANN R. 1965 Grundlagen und Anwendung der Atomabsorptionsspektroskopie in Flammen. *Z. Klin. Chem.* **31** 178.

*Herrmann R. and Alkemade C.Tr.J. 1960 *Flammenphotometrie*. Springer Verlag, Berlin. 394 pp.

Hewitt B.R. 1958 Spectrophotometric determination of total carbohydrate. *Nature, Lond.* **182** 246–247.

Holden A.V. and Marsden K. 1966 The examination of surface waters and sewage effluents for organochlorine pesticides. *J. Proc. Inst. Sew. Purif.* **3** 295–299.

Holden A.V. and Marsden K. 1969 Single-stage clean-up of animal tissue extracts for organochlorine residue analysis. *J. Chromat.* **44** 481–492.

Holmes R.W. 1943 Silver staining of nerve axons in paraffin section. *Anat. Rec.* **86** 157–186.

Holm-Hansen O., Lorenzen C., Holmes R.W. and Strickland J.D.H. 1965 Fluorimetric determination of chlorophyll. *J. Cons. perm. int. Explor. Mer* **30** 3–15.

Hulme A.C. and Narain R. 1931 The ferricyanide method for the determination of reducing sugars; a modification of the Hagedorn-Jensen-Hanes technique. *Biochem. J.* **25** 1051–1061.

*Hutchinson G.E. 1957 *A Treatise on Limnology* 3 vols. Wiley, New York. Vol. 1 Geography, physics and chemistry. 1015 pp.

*Jeffery P.G. and Kipping P.J. 1964 *Gas Analysis by Gas Chromatography*. Pergamon Press, Oxford. 216 pp.

*Jerlov N.G. 1976 *Marine Optics; 2nd rev. ed. of Optical Oceanography*. Elsevier, Amsterdam, New York. 232 pp.

Jerlov N.G. and Nygard K. 1969 Influence of solar elevation on attenuation of underwater irradiance. *Inst. Fysisk. Oceanografi. Univ. Copenhagen, Rep. nr.* **4** 9 pp.

Johnson D.L. 1971 Simultaneous determination of arsenate and phosphate in natural waters. *Environ. Sci. & Technol.* **5** 411–414.

*Jones A.G. 1960 *Analytical Chemistry. Some new techniques*. Butterworth, London. 268 pp.

Junk G.A., Richard J.J., Grieser M.D., Witiak D., Witiak J.L., Arguello M.D., Vick R., Svec H.J., Fritz J.S. and Calder G.V. 1974 Use of macroreticular resins in the analysis of water for trace organic contaminants. *J. Chromat.* **99** 745–762.

Jönsson E. 1966 The determination of Kjeldahl nitrogen in natural water. *Vattenhygien* **1** 10–14.

Kalle K. 1955/56 Fluoreszensmessungen in den Niderschlagwässern und in künstlichem Rauhreif. *Annln. Met., Hamburg* **7** 374–385.

*Karlgren L. 1962 *Vattenkemiska analysemetoder*. Limnologiska Institutionen, Uppsala.

Kay H. 1954 Eine Mikromethode zur chemischem Bestimmung des organisch gebundenen Kohlenstoffs im Meerwasser. *Kieler Meeresforsch.* **10** 26–36.

Keeler R.F. 1959 Colour reaction for certain amino acids, amines and proteins. *Science N.Y.* **129** 1617–1618.

Kerr J.R.W. 1960 The spectrophotometric determination of microgram amounts of calcium. *Analyst, Lond.* **85** 867–870.

Kiefer D.A. 1973a Fluorescence properties of natural phytoplankton populations. *Marine Biology* **22** 263–269.

Kiefer D.A. 1973b Chlorophyll *a* fluorescence in marine centric diatoms: responses of chloroplasts to light and nutrient stress. *Marine Biology* **23** 39–46.

Kleijn H.F.W. 1972 Partial Softening of Water. *H_2O* **5** 172.

Knowles G. and Lowden G.F. 1953 Methods for detecting the end point in the titration of iodine with thiosulphate. *Analyst, Lond.* **78** 159.

Kobayashi J. 1967 Silica in fresh water and estuaries. In Golterman and Clymo (Ed.) *q.v.* 41–55.

*Kolthoff I.M., Stenger V.A. and Belcher R. 1957 *Volumetric Analysis*. Interscience, New York. Vol. 3 *Titration Methods:* Oxidation-reduction reactions. 714 pp.

*Kolthoff I.M. and Sandell E.B. 1952 *Textbook of Quantitative Inorganic Analysis*; 3rd ed. MacMillan, New York. 759 pp.

*Kolthoff I.M. and Lingane J.J. 1952 *Polarography*; 2nd ed. 2 vols. Interscience. New York. Vol. 1: *Theoretical Principles, Instrumentation and Technique.* 420 pp. Vol. 2: *Inorganic Polarography, Organic Polarography, Biological Applications, Amperometric Titrations.* 990 pp.

Kovacs M.F. 1963 Thin-layer chromatography for chlorinated pesticide residue analysis. *J. Ass. off. agric. Chem.* **46** 884–893.

Krey J. 1951 Quantitative Bestimmung von Eiweiss im Plankton mittels der Biuretreaktion. *Kieler Meeresforsch.* **8** 16–29.

KREY J. and SZEKIELDA K-H. 1965 Bestimmung der organisch gebundenen Kohlenstoffs im Meerwasser mit einem neuen Gerät zur Analyse sehr kleiner Mengen CO_2. Z. *analyt. Chem.* **207** 338–346.

KRISHNAMURTY K.V., SHPIRT E. and REDDY M.M. 1976 Trace metal extraction of soils and sediments by nitric acid-hydrogen peroxide. *At. Absorp. Newsl.* **15** 68–70.

LARSON T.E. 1938 Properties and determination of methane in ground waters. *J. Am. Wat. Wks Ass.* **30** 1828.

*LEDERER E. and LEDERER M. 1957 *Chromatography; a Review of Principles and Applications;* 2nd ed. Elsevier Amsterdam, New York. 711 pp.

LEE G.F. 1967 Automatic methods. In Golterman and Clymo (Ed.) *q.v.* 169–179.

LEE G.F. and STUMM W. 1960 Determination of ferrous iron in the presence of ferric iron with bathophenanthroline. *J. Am. Wat. Ass.* **52** 1567–1574.

LEGLER Ch. 1972 Methoden der Sauerstoffbestimmung und ihre Bewertung. *Fortschr. Wasserchem. & Ihrer Grenzgeb.* **14** 27.

LEONI V., PUCCETTI G. and GRELLA A. 1975 Preliminary results on the use of Tenax® for the extraction of pesticides and polynuclear aromatic hydrocarbons from surface and drinking waters for analytical purposes. *J. Chromat.* **106** 119–124.

LEWIN R.A. 1961 Phytoflagellates and algae, in *Encyclopedia of Plant Physiology*. W. Ruhland (Ed.) Springer Verlag, Berlin. **14** 407–417.

LIDDICOAT M.I., TIBBITS S. and BUTLER E.I. 1976 The determination of ammonia in natural waters. *Wat. Res.* **10** 567–568.

*LINGANE J.J. 1958 *Electroanalytical Chemistry* 2nd. ed. Interscience, New York. 669 pp.

LIONNEL L.J. 1970 An automated method for the determination of boron in sewage, sewage effluents and river waters. *Analyst, Lond.* **95** 194–199.

LOFTUS M.E. and SELIGER H.H. 1975 Some limitations of the *in vivo* fluorescence technique. *Chesapeake Sci.* **16** 79–92.

LONG I.F. 1968 Instruments and techniques for measuring the microclimate of crops, in R.M. Wadsworth (Ed.) *The Measurement of Environmental Factors in Terrestrial Ecology*. 8th Symposium of the British Ecological Society. Blackwell, Oxford. 1–32.

LORENZEN C.J. 1966 A method for the continuous measurement of *in vivo* chlorophyll concentration. *Deep Sea Res.* **13** 223–227.

LUND J.W.G. and TALLING J.F. 1957 Botanical limnological methods with special reference to the algae. *Bot. Rev.* **23** 489–583.

*MCGLASHAN M.L. 1968 1st ed. 1971 2nd ed. *Physico-Chemical Quantities and Units.* Royal Institute of Chemistry, London. 116 pp.

MACIOLEK J.A. 1962 Limnological organic analyses by quantitative dichromate oxidation. Res. report nr. 60 of Bur. sport fish and wildlife. U.S. Fish and Wildlife service.

MACKERETH F.J.H. 1955 Rapid micro-estimation of the major anions of freshwater. *Proc. Soc. Wat. Treat. Exmn.* **4** 27–42.

*MACKERETH F.J.H., HERON J. and TALLING J.F. 1978 *Water Analysis: Some Methods for Limnologists.* Freshwater Biol. Ass. Sc. Publications.

MACKERETH F.J.H. 1964 An improved galvanic cell for determination of oxygen concentrations in fluids. *J. scient. Instrum* **41** 38–41.

MACKINNEY G. 1940 Criteria for purity of chlorophyll preparations. *J. biol. Chem.* **132** 91–109.

MACKINNEY G. 1941 Absorption of light by chlorophyll solutions. *J. biol. Chem.* **140** 315–322.

MADGWICK J.C. 1966 Chromatographic determination of chlorophylls in algal cultures and phytoplankton. *Deep Sea Res.* **13** 459–466.

MANCY K.H. and OKUN D.A. 1960 Automatic recording of dissolved oxygen in aqueous systems containing surface active agents. *Analyt. Chem.* **32** 108.

MANCY K.H., OKUN D.A. and REILLY C.N. 1962 A galvanic cell oxygen analyzer. *J. electroanal. Chem.* **4** 65.

MANCY K.H. and WESTGARTH W.C. 1962 A galvanic cell oxygen analyzer. *J. Wat. Pollut. Control Fed.* **34** 1037.

MARKER A.F.H. 1972 The use of acetone and methanol in the estimation of chlorophyll in the presence of phaeophytin. *Freshwater Biol.* **2** 361–385.

MARKER F.H. 1977 Some problems arising from the estimation of chlorophyll a and phaeophytin a in methanol. *Limnol. Oceanogr.* **22** 578–579.

MENZEL D.W. and CORWIN N. 1965 The measurement of total phosphorus in sea water based on the liberation of organically bound fractions by persulfate oxidation. *Limnol. Oceanogr.* **10** 280–282.

MENZEL D.W. and VACCARO R.F. 1964 The measurement of dissolved organic particulate carbon in sea water. *Limnol. Oceanogr.* **9** 138–142.

MILNER O.I. and ZAHNER R.J. 1960 Titration of traces of ammonia after Kjeldahl distillation. *Analyt. Chem.* **32** 294.

MITTELHOLZER E. 1970 Populationsdynamik und Production des Zooplanktons. *Hydrologie* **32** 106–139.

MONTEITH J.L. 1959 Solarimeter for field use. *J. scient. Instrum.* **36** 341–346.

*MONTEITH J.L. 1973 *Principles of Environmental Physics.* Arnold, London. 241 pp.

MONTGOMERY H.A.C. and COCKBURN A. 1964 Errors in sampling for dissolved oxygen. *Analyst, Lond.* **89** 679.

MONTGOMERY H.A.C. and QUARMBY C. 1966 The extraction of gases dissolved in water for analysis by gas chromatography. *Lab. Pract.* **15** 538–543.

MONTGOMERY H.A.C. and THOM N.S. 1962 The determination of low concentration of organic carbon in water. *Analyst, Lond.* **87** 689–697.

MONTGOMERY H.A.C., THOM N.S. and COCKBURN A. 1964 Determination of dissolved oxygen by the Winkler method and the solubility of oxygen in pure water and sea water. *J. appl. Chem., Lond.* **14** 280.

MOORE W.H. 1976 A battery-operated, flow-through thermistor thermometer. *Freshwater Biol.* **6** 409–411.

MORGAN J.J. and STUMM W. 1965 Analytical chemistry of aqueous manganese. *J. Am. Wat. Wks Ass.* **47** 107–119.

MORTIMER C.H. 1942 The exchange of dissolved substances between mud and water in lakes. *J. Ecol.* **30** 147–201.

MORTIMER C.H. 1953 A review of temperature measurement in limnology. *Mitt. int. Verein. theor. angew. Limnol.* nr. 1.

MORTIMER C.H. and MOORE W.H. 1953 (amendments 1970) The use of thermistors for the measurement of lake temperatures. *Mitt. int. Verein. theor. angew. Limnol.* nr. 2.

MURPHY J. and RILEY J.P. 1958 A single-solution method for the determination of soluble phosphate in sea water. *J. mar. biol. Ass. U.K.* **37** 9.

MURPHY J. and RILEY J.P. 1962 A modified single-solution method for the determination of phosphate in natural waters. *Analytica chim. Acta* **27** 31.

MUSTY P.R. and NICKLESS G. 1974a. Use of Amberlite XAD-4 for extraction and recovery of chlorinated insecticides and polychlorinated biphenyls from water. *J. Chromat.* **89** 185–190.

MUSTY P.R. and NICKLESS G. 1974b The extraction and recovery of chlorinated insecticides and polychlorinated biphenyls from water using porous polyurethane foams. *J. Chromat.* **100** 83-93.

MUZZARELLI R.A. and ROCHETTI R. 1973 The determination of molybdenum in sea water by hot graphite atomic absorption after concentration on p-aminobenzylcellulose or chitosan. *Analytica chim. Acta* **64** 371–379.

NICOLSON N.J. 1966 The colorimetric determination of iron in water. *Proc. Soc. Wat. Treat. Exam.* **15** 157–158.

NIEPCE de SAINT-VICTOR et CORVISART L. 1859 De la fécule végétale et animale sous le rapport de l'influence transformatrice qu'exerce sur elle la lumière solaire; de la dextrine, du sucre de canne, de l'acide oxalique sous le même rapport; de quelques substances qui annihilent ou accroissent cette action solaire. *C.r. Acad. Sci., Paris.* **49** 368–371.

O'CONNOR J.T. and RENN C.E. 1963 Evaluation of procedure for the determination of zinc. *J. Am. Wat. Wks Ass.* **55** 631–638.

O'CONNOR J.T., KOMOLRIT K. and ENGLEBRECHT R.S. 1965 Evaluation of the orthophenanthroline method for the ferrous iron determination. *J. Am. Wat. Wks Ass.* **57** 926–933.

OLSEN S. 1967 Recent trends in the determination of orthophosphate in water. In Golterman and Clymo (Ed.) *q. v.* 63–105.

OTSUKA H. and MORIMURA Y. 1966 Change of fatty acids composition of *Chlorella ellipsoidea* during its cell cycle. *Pl. Cell Physiol., Tokyo* **7** 663–670.

PARK K. 1965 Gas-chromatographic determination of dissolved oxygen, nitrogen, and total carbon dioxide in sea water. *J. oceanogr. Soc. Japan* **21** 28–29.

PARK K. and CATALFOMO M. 1964 Gas-chromatographic determination of dissolved oxygen in seawater using argon as carrier gas. *Deep Sea Res.* **11** 917–920.

PARSONS T.R. and STRICKLAND J.D.H. 1963 Discussion of spectrophotometric determination of marine-plant pigments, with revised equations for ascertaining chlorophylls and carotenoids. *J. mar. Res.* **21** 155–163.

**Pesticide Analytical Manual, Vol. I.* 1968 Food and Drug Administration, Washington D.C.

POMEROY R. and KIRSCHMAN H.D. 1945 Determination of dissolved oxygen; proposed modification of the Winkler method. *Ind. Engng. Chem. analyt. Edn* **17** 715–716.

POWELL M.C. and HEATH O.V.S. 1964 A simple and inexpensive integrating photometer. *J. exp. Bot.* **15** 187–191.

*PRICE W.J. 1974 *Analytical Atomic Absorption Spectrometry*. Heyden, London. 251 pp.

PROCTOR C.M. and HOOD D.W. 1954 Determination of inorganic phosphate in sea water by an iso-butanol extraction procedure. *J. mar. Res.* **13** 122–132.

*RANDERATH K. 1966 *Thin Layer Chromatography*. Academic Press, New York, London. 285 pp.

RAY N.H. 1954 Gas chromatography. II The separation and analysis of gas mixtures by chromatographic methods. *J. appl. Chem., Lond.* **4** 82–85.

REBSDORF A. 1966 Evaluation of some modification of the Winkler method for the determination of oxygen in natural water. *Verh. internat. Verein. theor. angew. Limnol.* **16** 459–464.

*REYNOLDS R.J. and ALDOUS K. 1969 *Atomic Absorption Spectroscopy*. Griffin, London. 210 pp.

RICHARDS F.A. and THOMPSON T.G. 1952 The estimation and characterization of plankton populations by pigment analyses. II A spectrophotometric method for the estimation of plankton pigments. *J. mar. Res.* **11** 156–172.

RILEY J.P. and SINHASENI P. 1958 The determination of copper in sea water, silicate rocks and biological materials. *Analyst, Lond.* **83** 299–304.

*RILEY J.P. and SKIRROW G. (Ed.) 1965 *Chemical Oceanography*. Academic Press, New York. Vol. 1 712 pp. Vol. 2 508 pp.

RILEY J.P. and WILSON T.R.S. 1967 The pigments of some marine phytoplankton species. *J. mar. biol. Ass. U.K.* **47** 351–362.

RODHE W. 1949 The ionic composition of lake waters. *Verh. internat. Verein. theor. angew. Limnol.* **10** 377–386.

ROSKAM R. Th. and LANGEN D. de 1963 A compleximetric method for the determination of dissolved oxygen in water. *Analytica chim. Acta* **28** 78–81.

ROSSUM J.R., VILLARRUZ P.A. and WADE J.A.A. 1950 A new method for determining methane in water. *J. Am. Wat. Wks Ass.* **42** 413–415.

*SANDELL E.B. 1959 *Colorimetric Determination of Traces of Metals*. Vol. 3 Chemical Analysis. 3rd ed. Interscience, New York. 1032 pp.

SCHEINER D. 1976 Determination of ammonia and Kjeldahl nitrogen by indophenol method. *Wat. Res.* **10** 31–36.

SCHINDLER J.E. and HONICK K.R. 1971 Oxidation-reduction determinations at the mud-water interface. *Limnol. Oceanogr.* **16** 837

SCHMITT Cl. 1955 Contribution a l'étude du système cheaux-carbonate de calcium-bicarbonate de calcium-acide carbonique-eau. *Annls. Ec. natn. sup. mecanique*, Nantes.

SCOR working group 15 1965 Report of the first meeting of the joint group of experts in photosynthetic radiant energy. *Unesco Tech. Pap. Mar. Sci.* **2** 1–5.

SCOR working group 17 1966 *Determination of photosynthetic pigments in sea water*. Unesco, Paris. 69 pp.

SHAPIRO J. 1961 Freezing-out, a safe technique for concentration of dilute solutions. *Science, N.Y.* **133** 2063–2064.

SHAPIRO J. 1965 On the measurement of ferrous iron in natural waters. *Limnol. Oceanogr.* **11** 293–298.

SHAPIRO J. 1967a Yellow organic acids of lake water: differences in their composition and behaviour. In Golterman and Clymo (Ed.) *q.v.* 202–216.

SHAPIRO J. 1967b Iron available to algae. In Golterman and Clymo (Ed.) *q.v.* 219–228.

SKOPINTSEV B.A. 1960 Trudy morsk gidrofiz. Inst. 19 3; translated in *Soviet Oceanography* 1963 Series nr. **3** 1–14.

SLYKE D.D. van, DILLON R.T. van, MACFADGEN D.A. and HAMILTON P. 1941 Gasometric determination of carboxyl groups in free amino acids. *J. biol. Chem.* **141** 627–669.

SMITH R.C. 1969 An underwater spectral irradiance collector. *J. mar. Res.* **27** 341–351.

SMITH R.G. Jr. 1974 Improved ion-exchange technique for the concentration of manganese from sea water. *Analyt. Chem.* **46** 607–608.

SPENCE D.H.N. 1975 Light and plant response in fresh water. In Evans, Rackham and Bainbridge (Ed.) *q.v.* 93–134.

*SQUIRREL D.C.M. 1964 *Automatic Methods in Volumetric Analysis*. Hilger and Watts, London. 201 pp.

STABEL H-H. 1977 On the problem of the determination of dissolved carbohydrates with the anthrone method. *Arch. Hydrobiol.* **80** 216–226.

*STAHL E. (Ed.) 1970 *Thin-layer Chromatography. A Laboratory Handbook*. Springer-Verlag, Berlin. 1041 pp.

STAINTON M.P. 1973 A syringe gas-stripping procedure for gas chromatographic determination of dissolved inorganic and organic carbon in freshwater and carbonates in sediments. *J. Fish. Res. Bd Can.* **30** 1441–1445

**Standard Methods for the Examination of Water and Wastewater, including Bottom Sediments and Sludge*; 12th ed. 1965 Am. publ. Hlth Ass. New York. 769 pp.

STANGENBERG M. 1959 Der biochemische Sauerstoffbedarf des Seewassers. *Memorie Ist. ital. Idrobiol.* **11** 185–211.

STEPHENS K. 1963 Determination of low phosphate concentrations in lake and marine waters. *Limnol. Oceanogr.* **8** 361–362.

STRAŠKRABOVÁ-PROKEŠOVÁ V. 1966 Oxidation of organic substances in the water of the reservoirs Slapy and Klicava, in *Hydrobiological studies*. J. Hrbáček (Ed). 85–111 Acad. Publ. House, Prague.

*STRICKLAND J.D.H. and PARSONS T.R. 1968 *A Practical Handbook of Seawater Analyses*. Bull. Fish. Res. Bd Can. 167.

*STUMM W. and MORGAN J.J. 1970 *Aquatic Chemistry*. Wiley Interscience. New York. 583 pp.

SWINNERTON J.W., LINNENBOM V.J. and CHEEK C.H. 1962a Determination of dissolved gases in aqueous solution by gas chromatography. *Analyt. Chem.* **34** 483–485.

SWINNERTON J.W., LINNENBOM V.J. and CHEEK C.H. 1962b Revised sampling procedure for determination of dissolved gases in solution by gas chromatography. *Analyt. Chem.* **34** 1509.

SWINNERTON J.W., LINNENBOM V.J. and CHEEK C.H. 1964 Determination of argon and oxygen by gas chromatography. *Analyt. Chem.* **36** 1669–1671.

SZEICZ G. 1966 Field measurements of energy in the 0.4–0.7 micron range. In R. Bainbridge, G.C. Evans and O. Rackham (Ed.) *Light as an Ecological Factor*. 6th Symposium of the British Ecological Society. Blackwell; Oxford. pp. 41–42.

SZEICZ G.C. 1975 Field measurements of energy in the 0.4–0.7 micron range II. In Evans, Rackham

SZEKIELDA K-H. and KREY J. 1965 Die Bestimmung des partikulären organisch gebundenen Kohlenstoffs im Meerwasser mit einer neuen Schnellmethode. *Mikrochim. Acta* **1** 149–159.

TALLING J.F. 1973 The application of some electrochemical methods to the measurement of photosynthesis and respiration in fresh waters. *Freshwater Biol.* **3** 335–362.

TALLING J.F. and DRIVER D. 1963 Some problems in the estimation of chlorophyll-*a* in phytoplankton. Proc. conference on primary productivity measurements, marine and fresh-water; held at Univ. Hawaii 1961. U.S. Atomic Energy Comm. TID–7633 142–146.

TARAS M. 1948 Photometric determination of magnesium in water with Brilliant Yellow. *Analyt. Chem.* **20** 1156–1158.

TILLMANS J. and HEUBLEIN O. 1912 Uber die Kohlensäuren Kalk angreifende Kohlensäure der natürlichen Wässer. Gesundheits-ingenieur. **35** 669–677.

TYLER J.E. 1973 Lux vs. quanta. *Limnol. Oceanogr.* **18** 810.

UPHOFF G.D. and HERGENRADER G.L. 1976 A portable quantum meter-spectroradiometer for use in aquatic studies. *Freshwater Biol.* **6** 215–219.

UTHE J.F., REINKE J. and GESSER H. 1972 Extraction of organochlorine pesticides from water by porous polyurethane coated with selective absorbent. *Envir. Letters* **3** 117–135.

VALLENTYNE J. 1967 Pheromones and related substances. In Golterman and Clymo (Ed.) *q.v.* 252–254.

VERNON L.P. 1960 Spectrophotometric determinations of chlorophylls and phaeophytins in plant extracts. *Analyt. Chem.* **32** 1144–1150.

VIJVERBERG J. and FRANK Th.H. 1976 The chemical composition and energy contents of copepods and cladocerans in relation to their size. *Freshwater Biol.* **6** 333–345.

*VOGEL A.I. 1961 *A textbook of Quantitative Inorganic Analysis*. 3rd ed. Longmans, London. 1216 pp.

*VOLLENWEIDER R.A. (Ed.) 1971 *Primary Production in Aquatic Environments*. Blackwell Scientific Publications, Oxford. 225 pp.

WALPOLE G.St. 1914 Hydrogen potentials of mixtures of acetic acid and sodium acetate. *J. chem. Soc.* **105** 2501–2529.

WALSH A. 1965 Some recent advances in atomic-absorption spectroscopy. *Coll. Spectr. Internat.* 12.

WATT W.D. 1965 A convenient apparatus for *in situ* primary production studies. *Limnol. Oceanogr.* **10** 298–300.

WEBER W.J. and STUMM W. 1963 Mechanism of hydrogen ion buffering in natural waters. *J. Am. Wat. Wks Ass.* **55** 1558–1578.

*WEISSBERGER A. and ROSSITER B.W. (Ed.) 1971 *Physical Methods of Chemistry*. Wiley Interscience, New York, London.
Part I Components of scientific instruments, automatic recording and control, computers in chemical research. 448 pp.
Part II A Electrochemical methods. 723 pp.

WESTLAKE D.F. 1965 Some problems in the measurement of radiation under water; a review. *Photochem. & Photobiol.* **4** 849–868.

WESTLAKE D.F. 1966 The light climate for plants in rivers. In R. Bainbridge, G.C. Evans and O. Rackham (Ed.) *Light as an Ecological Factor*. 6th Symposium of the British Ecological Society. Blackwell, Oxford. pp, 99–119.

WESTLAKE D.F. and DAWSON F.H. 1975. In Evans, Rackham and Bainbridge (Ed.) *q.v.* 1–25.

*WHITFIELD M. 1971 *Ion Selective Electrodes foı the Analysis of Natural Waters*. Australian Mar. Sci. Assoc. Sydney. 130 pp. (obtainable from: AMSA, c/o Hydrographic Office, R.A.N., Garden Island, N.S.W. 2000, Australia).

WILCOX L.V. 1950 Electrical conductivity. *J. Am. Wat. Wks Ass.* **42** 775–776.

*WILSON A.L. 1974 *The Chemical Analysis of Water*. Society for Analytical Chemistry, London. 188 pp.

·WOOD E.D., ARMSTRONG F.A.J. and RICHARDS F.A. 1967 Determination of nitrate in sea water by cadmium-copper reduction to nitrite. *J. mar. biol. Ass. U.K.* **47** 23–31.

YEMM E.W. and WILLIS A.J. 1954 The estimation of carbohydrates in plant extracts by anthrone. *Biochem. J.* **57** 508–514.

YENTSCH C.S. and MENZEL D.W. 1963 A method for the determination of phytoplankton chlorophyll and phaeophytin by fluorescence. *Deep Sea Res.* **10** 221–231.

YOUNG P.A. 1949 Personal error in penicillin assay. *Br. J. Pharmac. Chemother.* **4** 366–372.

ZOBELL Cl. E. and BROWN B.F. 1944 Studies on the chemical preservation of water samples. *J. mar. Res.* **5** 178–184.

ZÖLLNER N. and KIRSCH K. 1962 Über die quantitative Bestimmung von Lipoiden (Mikromethode) mittels der vielen natürlichen Lipoiden (allen bekannten Plasmalipoiden) gemeinsamen sulphophosphovanillin-Reaktion. *Z. Gesamte Exp. Med.* **135** 545–561.

ZSCHEILE F.P. 1934a A quantitative spectro-photoelectric analytical method applied to solutions of chlorophylls *a* and *b*. *J. phys. Colloid Chem.* **38** 95–102.

ZSCHEILE F.P. 1934b An improved method for the purification of chlorophylls *a* and *b*; the quantitative measurement of their absorption spectra; evidence for the existence of a third component of chlorophyll. *Bot. Gaz.* **95** 529–562.

ZSCHEILE F.P., COMAR C.L. and MACKINNEY G. 1942 Interlaboratory comparison of absorption spectra by the photolectric spectrophotometric method. Determinations on chlorophyll and Weigert's solutions. *Plant. Physiol., Lancaster.* **17** 666–670.

INDEX

This index is intended to supplement the Table of Contents, pp v-xiii

Conversion Table of Temperature F° to C°

$$\left(\frac{F - 32}{9} = \frac{C}{5}\right)$$

F°	0	1	2	3	4	5	6	7	8	9
30	−1·1	−0·6	0·0	0·6	1·1	1·7	2·2	2·8	3·3	3·9
40	4·4	5·0	5·6	6·1	6·7	7·2	7·8	8·3	8·9	9·4
50	10·0	10·6	11·1	11·7	12·2	12·8	13·3	13·9	14·4	15·0
60	15·6	16·1	16·7	17·2	17·8	18·3	18·9	19·4	20·0	20·6
70	21·1	21·7	22·2	22·8	23·3	23·9	24·4	25·0	25·6	26·1
80	26·7	27·2	27·8	28·3	28·9	29·4	30·0	30·6	31·1	31·7
90	32·2	32·8	33·3	33·9	34·4	35·0	35·6	36·1	36·7	37·2